Understanding Advanced Chemistry Through Problem Solving

(Revised Edition)

The Learner's Approach **Volume 1**

Understanding Advanced Chemistry Through Problem Solving

(Revised Edition)

The Learner's Approach **Volume 1**

Kim Seng Chan

BSc (Hons), PhD, PGDE (Sec), MEd, MA (Ed Mgt), MEd (G Ed), MEd (Dev Psy)

Jeanne Tan

BSc (Hons), PGDE (Sec), MEd (LST)

EW JERSEY • LONDON • SINGAPORE • BEIJING • SHANGHAI • HONG KONG • TAIPEI • CHENNAI • TOKYO

Published by

World Scientific Publishing Co. Pte. Ltd.
5 Toh Tuck Link, Singapore 596224
USA office: 27 Warren Street, Suite 401-402, Hackensack, NJ 07601
UK office: 57 Shelton Street, Covent Garden, London WC2H 9HE

British Library Cataloguing-in-Publication Data
A catalogue record for this book is available from the British Library.

UNDERSTANDING ADVANCED CHEMISTRY THROUGH PROBLEM SOLVING
The Learner's Approach
(In 2 Volumes)
Revised Edition

ISBN 978-981-12-8968-2 (set_hardcover)
ISBN 978-981-12-8979-8 (set_paperback)
ISBN 978-981-12-8270-6 (vol. 1_hardcover)
ISBN 978-981-12-8179-2 (vol. 1_paperback)
ISBN 978-981-12-8180-8 (vol. 1_ebook for institutions)
ISBN 978-981-12-8181-5 (vol. 1_ebook for individuals)
ISBN 978-981-12-8271-3 (vol. 2_hardcover)
ISBN 978-981-12-8182-2 (vol. 2_paperback)
ISBN 978-981-12-8183-9 (vol. 2_ebook for institutions)
ISBN 978-981-12-8184-6 (vol. 2_ebook for individuals)

For any available supplementary material, please visit
https://www.worldscientific.com/worldscibooks/10.1142/13555#t=suppl

Typeset by Stallion Press
Email: enquiries@stallionpress.com

PREFACE

When a major examination approaches, students would start going around in search for guidebooks that can help them to consolidate the important concepts that are necessary to meet the requirements of the assessments in the shortest amount of time. But unfortunately, most guidebooks are of the expository and non-refutational type, presenting facts rather than explaining them. In addition, the links between concepts are often not made explicit and presupposes that learners would be able to make the necessary integration with the multitude of concepts that they have come across in their few years of chemical education, forgetting that some learners may lack the prior knowledge and metacognitive skills to do it meaningfully. Hence, learners would at most be able to reproduce the information that is structured and organized by the guidebook writer, but not able to construct a meaningful conceptual mental model for oneself. Hence, the learners would not be able to apply what they should know fluidly across different contextual questions that appear in the major examination.

This current book is a continuation of our previous two books — *Understanding Advanced Physical Inorganic Chemistry* and *Understanding Advanced Organic and Analytical Chemistry*, retaining the main refutational characteristics of the two books by strategically planting think-aloud questions to promote conceptual understanding, knowledge construction, reinforcement of important concepts, and discourse opportunities. It is hoped that these essential questions would make learners aware of the possible conflict between their prior knowledge, which may be counterintuitive or misleading, with those presented in the text, and hence in the process, make the necessary conceptual changes. In essence, we are trying to effect metaconceptual awareness — awareness of the theoretical nature of one's thinking — while learners are trying to master the essential chemistry concepts and learn about their applications in problem solving. We hope that by pointing out the differences between possible misconceptions and the actual chemistry content, we can promote metaconceptual awareness and thus assist the learner to construct a meaningful conceptual model of understanding to meet the necessary assessment criteria. We want our learners to not only know what they know, but at the same time, have a sense of how they know what they know and how their new learning is interrelated within the discipline. This would enable the learners to better appreciate and fluidly apply what they have learned in whatever novel questions that they come across in the major examinations.

Lastly, the substance in this book would be both informative and challenging to the practices of teachers. This book would certainly illuminate the teaching of all chemistry teachers who strongly believe in teaching chemistry in a meaningful and integrative approach, from the learners' perspective. The integrated questions that are being used as problem-solving tools would certainly prove useful to students in helping them to revise fundamental concepts that they have learned from previous chapters, and also perceive the importance and relevancy in the application to their current learning. Collectively, this book offers a vision of understanding and applying chemistry meaningfully and fundamentally from the learners' approach and to fellow chemistry teachers, we hope that it would help you develop a greater insight into what makes you tick, explain, enthuse, and develop in the course of your teaching.

Kim Seng Chan
BSc (Hons), PhD, PDGE (Sec), MEd, MA (Ed Mgt), MEd (G Ed), MEd (Dev Psy)

Jeanne Tan
BSc (Hons), PDGE (Sec), MEd (LST)

ACKNOWLEDGEMENTS

We would like to express our sincere thanks to the staff at World Scientific Publishing Co. Pte. Ltd. for the care and attention which they have given to this book, particularly our editors Lim Sook Cheng, Sandhya Devi and Stallion Press.

We also wish to take this opportunity to express our gratitude to Yeo Yam Khoon of Raffles Institution (Year 5-6) who has been and will always be an inspiration for his unwavering commitment to the teaching of chemistry.

Special thanks to Tham Zisheng of Raffles Institution (Year 1-4), Ms Soh Xiao Fen of St. Andrew Junior College, Mdm Toh Chui Hoon of Victoria Junior College, Dr Chok Yew Keong of Eunoia Junior College and William Wong Yu Kai of Victoria Junior College for some of their invaluable suggestions and mind provoking discussions, in the past and present.

Special thanks go to all our students who have made our teaching of chemistry fruitful and interesting. We have learnt a lot from them just as they have learnt some good chemistry from us.

Finally, we thank our families for their wholehearted support and understanding throughout the period of writing this book. We would like to share with all the passionate learners of chemistry two important quotes from the *Analects of Confucius*:

學而時習之，不亦悅乎？**(Isn't it a pleasure to learn and practice what is learned time and again?)**

學而不思則罔，思而不學則殆 **(Learning without thinking leads to confusion, thinking without learning results in wasted effort)**

CONTENTS

PART I

PHYSICAL CHEMISTRY

CHAPTER 1

ATOMIC STRUCTURE AND THE PERIODIC TABLE

1. Naturally occurring boron consists of two isotopes, ^{10}B and ^{11}B, having abundances of 19.7% and 80.3%, respectively.
   (a) Explain the terms *isotope* and *relative isotopic mass*.

Explanation for *isotope*:

An element may consist of two or more atoms which have the same number of protons, also known as the atomic number, but different number of neutrons. These atoms are known as isotopes.

Do you know?

— The number of protons is equivalent to the number of electrons for isotopes. Hence, an isotope is an electrically neutral species.
— Isotopes have the same chemical properties as they have identical electronic configuration, and chemical reaction only involves the movement of electrons. The nucleus is intact during a chemical reaction.
— Isotopes have different physical properties such as melting point and boiling point. Isn't it more difficult to vaporize a heavier atom from its liquid state because of its heavier mass?

Explanation for *relative isotopic mass*:

Relative isotopic mass is the mass of one atom of the isotope of an element relative to 1/12 of the mass of one atom of ^{12}C.

Do you know?

— Relative isotopic mass is a dimensionless quantity as it is measured relatively to the mass of one atom of ^{12}C.
— The difference in the masses of isotopes arises because of the difference in the number of neutrons. The sum of protons and neutrons is known as the mass number or nucleon number.
— The value is similar to the mass number of the isotope, which is sum of protons and neutrons. This is because the bulk mass of an atom is attributed to what is present in the nucleus.
— The mass of an isotope seems to be a whole number. This is not true because in reality, the actual relative isotopic mass is less than the mass of all nucleons added up. This phenomenon is known as mass defect. The difference in the masses, which is less than 1%, arises because part of the mass has been converted into binding energy according to $E = mc^2$, which is necessary to hold the nucleons together in the nucleus.
— Relative isotopic mass can be determined using mass spectrometry.

Q Why is ^{12}C used as the reference standard?

A: Avogadro's number or Avogadro's Constant (N_A) is a common constant used in both chemistry and physics. Avogadro's number is formally defined as the number of carbon-12 atoms in 12 g of carbon-12, which is approximately 6.022×10^{23}. Historically, carbon-12 was chosen as the reference standard because its atomic mass could be measured with particular accuracy. In addition, carbon-12 is commonly present in many substances.

(b) Calculate the relative atomic mass of naturally occurring boron.

Explanation:

The relative atomic mass (A_r) of an element is dependent on (i) whether it has more than one isotope, and (ii) the composition of the various isotopes. Hence, to determine the relative atomic mass, we need the following formula:

$$A_r = \Sigma(\text{Percentage composition} \times \text{Relative isotopic mass})$$

Thus, A_r of boron = $(0.197 \times 10 + 0.803 \times 11) = 10.8$.

Do you know?

— Since the dimensionless relative isotopic mass is used in the calculation of the relative atomic mass, the latter is also a dimensionless quantity. Anyway, it is also a dimensionless quantity simply because it is a relative comparison to another quantity that has the same dimension, which is mass here.

— Based on the formula used for the calculation, relative atomic mass is a weighted average quantity.

— The greater the contribution of a particular isotope, the closer the relative atomic mass is to the value of the relative isotopic mass contributed by this particular isotope.

— The relative composition of the isotopes in the element would be the same as that present in the compound. For example, you would find 19.7% of ^{10}B and 80.3% of ^{11}B in a molecular compound that contains one boron atom.

2. (a) (i) Give the electronic configuration of an atom of the isotope of calcium, $^{45}_{20}Ca$.

Explanation:

The following guideline is used to craft the electronic configuration of an atom:

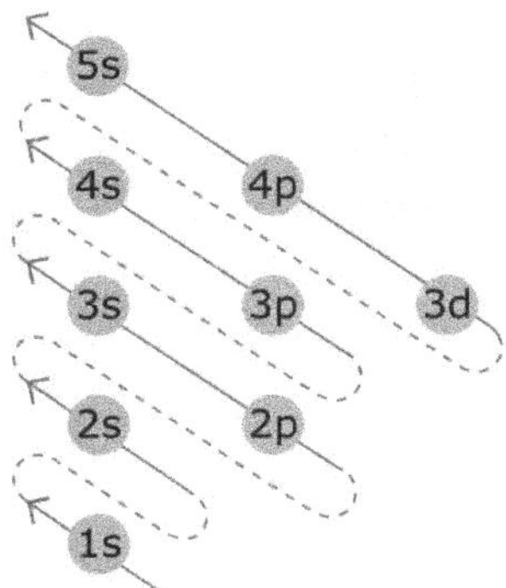

The electronic configuration of $^{45}_{20}\text{Ca}$ is $1s^2 2s^2 2p^6 3s^2 3p^6 4s^2$ since Ca has 20 electrons as indicated by the atomic number.

Do you know?

— Electron occupies the lowest energy orbital first — Aufbau's Principle.

— Each orbital can maximally accommodate two electrons with opposite spin, i.e., $+\frac{1}{2}$ and $-\frac{1}{2}$ — Pauli Exclusion Principle.

— When filling a set of degenerate orbitals (referring to orbitals in the same subshell and they are of the same energy), we need to ensure that each orbital is half full before completely filling up any one of the half full orbital — Hund's Rule of Multiplicity.

— The electronic configuration is always written in the order of increasing Principal Quantum Number. This also indicates the order of increasing energy level of the various subshells. For example, the electronic configuration of $_{25}\text{Mn}$ is written as $1s^2 2s^2 2p^6 3s^2 3p^6 3d^5 4s^2$ although the $4s$ orbital is initially filled first before filling up the $3d$ orbital.

(*Continued*)

(*Continued*)

- To obtain the electronic configuration of a cation or anion, craft up the electronic configuration of the neutral atom first. For the formation of cation, remove electrons from the highest energy subshell which contains electrons, i.e., the valence shell. For the formation of anion, put in electrons into the valence shell.
- Two species with the same electronic configuration is known as isoelectronic species.
- Knowing the electronic configuration is important as the removal of electrons from different orbitals with different energies would affect the rate of reaction, the energetics of the reaction, etc.

(ii) Gives the names and numbers of each type of nucleons present in the isotope.

Explanation:

There are 20 protons and 25 neutrons (45 − 20 = 25).

Do you know?

- A nucleon is either a proton or neutron.
- Left subscript of the elemental symbol represents the atomic number, while left superscript indicates the mass number, which is the sum of protons and neutrons.
- Right subscript indicates the number of the atom in the species, for example $(COOH)_2$ represents two units of –COOH bonded together while $Al_2(SO_4)_3$ represents 2 Al^{3+} and 3 SO_4^{2-}.
- Right superscript represents the charge of the species.

(iii) State one reason why the information in *(a)(i)* is usually more useful to chemists than that in *(a)(ii)*.

Explanation:

The electronic configuration is more important than the proton number because in a chemical reaction, it is the electrons that are transferred between different atoms; atom either gains, loses, or shares electrons. The nucleus remains intact during the chemical reaction!

Does that mean that the number of protons will not affect the chemical reaction at all?

A: The number of protons would determine the electrostatic attractive force the nucleus has for all its electrons. In an atom, the protons attract on all the electrons. The farther the electron is from the nucleus, the weaker is the electrostatic attractive force ($F \propto 1/r^2$). At the same time, there is also inter-electronic repulsion between the various electrons. The inter-electronic repulsion between the valence electrons and the inner core electrons is known as the shielding effect. The shielding effect opposes the attractive force the nucleus act on the valence electrons. Hence, the net electrostatic attractive that is actually acting on the valence electrons is known as the effective nuclear charge (ENC). The greater the ENC, (i) the higher the ionization energy (I.E.) (ii) the greater the electronegativity value, which refers to the ability of the atom to polarize shared electrons, (iii) the lower the reducing power, (iv) the greater the oxidizing power and (v) the smaller the atomic radius.

Now, since a chemical reaction only involves the transfer of electrons between different atoms, obviously the ENC would affect the rate of the reaction and hence the overall energetics of the reaction. For example, an atom which has a weaker ENC on its valence electrons would likely lose the electrons rather than gain electrons. And if this atom is involved in the sharing electrons, the attractive force on the shared electrons would also be weaker. So, how can we say that the number of protons would not affect the type of chemical reaction and the type of chemical product that forms?

(c) $^{45}_{20}Ca$ is radioactive, decaying by giving off electrons. State and explain what would happen to the electrons if they were passed through an electric field.

Explanation:

Electrons, being negatively charged particles, would be attracted to the positive plate of the electric field.

Do you know?

— Positively charged particles e.g., a proton, would be attracted to the negative plate.
— The angle of deflection $\propto \frac{\text{charge}}{\text{mass}}$.
— The radius of deflection $\propto \frac{\text{mass}}{\text{charge}}$
— The concept of deflection is used in mass spectrometry.

(d) (i) For the following species, sketch a graph of ionic size against their respective atomic numbers.

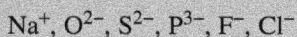

Na^+, O^{2-}, S^{2-}, P^{3-}, F^-, Cl^-

Explanation:

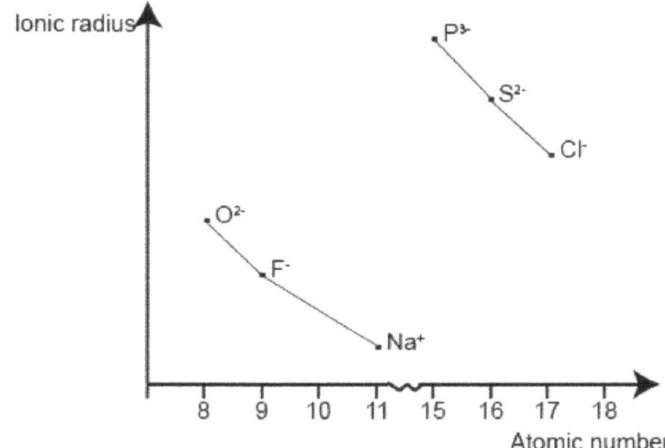

(ii) Explain the trends shown in the graph sketched in *(c)(i)*.

Explanation:

Electronic configuration of the various species:

Na^+	$1s^22s^22p^6$ (11 protons)	Cl^-	$1s^22s^22p^63s^23p^6$ (17 protons)
F^-	$1s^22s^22p^6$ (9 protons)	S^{2-}	$1s^22s^22p^63s^23p^6$ (16 protons)
O^{2-}	$1s^22s^22p^6$ (8 protons)	P^{3-}	$1s^22s^22p^63s^23p^6$ (15 protons)

Since Na^+, F^-, and O^{2-} are isoelectronic, they have the same amount of inter-electronic repulsion. But the decreasing nuclear charge from Na^+ to F^- to O^{2-} causes the net attractive force acting on the electrons to decrease. Hence, the ionic radius increases from Na^+ to F^- to O^{2-}.

Similarly, the ionic radius increases from Cl^- to S^{2-} to P^{3-}. The ionic radii for Cl^-, S^{2-}, and P^{3-} are much greater than Na^+, F^-, and O^{2-}, because the former group of ions has an extra principal quantum shell of electrons as compared to the latter group, which means the outermost electronic shell is farther from the nucleus. Hence, overall, the former group of ions (Cl^-, S^{2-} and P^{3-}) has a much bigger ionic radii than Na^+, F^- and O^{2-}.

Do you know?

— The size of a species, be it a neutral atom, a cation, or an anion, is a consequence of the resultant interaction between the inter-electronic repulsion of all the electrons and the nuclear charge acting on these electrons.

— For two isoelectronic species, they have same amount of inter-electronic repulsion, but the one that has a greater nuclear charge will result in a smaller atomic/ionic size.

— Similarly, for two cations that have the same nuclear charge but different number of electrons, e.g., Ca^+ and Ca^{2+}, the one that has more electrons would have greater inter-electronic repulsion, thus greater ionic size.

(*Continued*)

(Continued)

— The following shows the trends for the atomic and ionic radii of some elements:

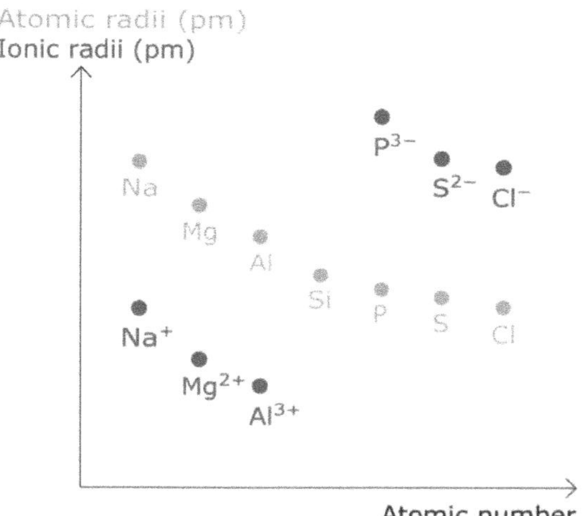

Do you notice that cationic radius is smaller than atomic radius, while anionic radius is greater than atomic radius?

3. Given the first four ionization energies, in kJ mol^{-1}, of gallium and cobalt:

	1st	2nd	3rd	4th
Gallium	577	1980	2960	6190
Cobalt	757	1640	3230	5100

(a) Provide an explanation for the increases in the successive ionization energies.

Explanation:

Successive ionization energies increase because as more electrons are being removed, the number of electrons decreases but the nuclear charge remains the same. Hence, the inter-electronic repulsion decreases. This results in a stronger electrostatic attractive force acting on the remaining electrons, which thus needs more energy to be removed.

(b) With reference to the electronic configuration, explain why a discontinuity is present in gallium but not in cobalt.

Explanation:

Ga $1s^2 2s^2 2p^6 3s^2 3p^6 3d^{10} 4s^2 4p^1$
Co $1s^2 2s^2 2p^6 3s^2 3p^6 3d^7 4s^2$

The discontinuity in Ga occurs at the 4th I.E. This is because the fourth electron to be removed from Ga comes from a 3*d* subshell which has a lower energy than the 4*s* and 4*p* subshells. In addition, the removal of the fourth electron would destroy the even symmetrical distribution of electron density in the Ga^{3+} ion.

As for Co, the difference in energy levels between the 3*d* and 4*s* subshells is not too great, thus there is no sudden jump in the I.E. The more significant change occurs at the 3rd I.E., since the third electron to be removed come from a lower-energy principal quantum shell.

Do you know?

— As the number of protons increases, the energy level of each of the subshell decreases. For example, the 2*s* orbital of boron is at a higher energy level than the 2*s* orbital of oxygen atom.
— At the point when the atomic number is 19, i.e., for potassium, the energy level of the 4*s* is in fact lower than the 3*d*. Thus, according to Aufbau's principle, the next electron would need to occupy the 4*s* subshell. This accounts for why potassium's electronic configuration is $1s^2 2s^2 2p^6 3s^2 3p^6 4s^1$.

(*Continued*)

(Continued)

— Subsequently, filling up of the 3*d* subshell causes the repulsion of the electrons in the 4*s* subshell as the 3*d* subshell is relatively closer to the nucleus than the 4*s* subshell. This further raised the energy level of the 4*s* subshell. But still, the 3*d* and 4*s* subshells still have relatively similar energy levels. And this accounts for why transition elements can form compounds with different oxidation states. But take note that the 4*p* subshell's energy is still relatively far away from the 4*s* subshell.

— Toward the end of the *d*-block, the energy gap between the 3*d* and 4*s* subshells have increased to an extent that after the removal of electrons from the 4*s* subshell, further removal of electron from the 3*d* subshell causes a more significant increase in I.E.

(c) Naturally occurring gallium consists of two isotopes, ^{69}Ga and ^{71}Ga. Calculate the percentage abundance of each isotope.

Explanation:

Let the compositional amount of ^{69}Ga be a. Hence, compositional amount of ^{71}Ga is $(1 - a)$.

A_r of gallium = $\{a \times 69 + (1 - a) \times 71\} = 69.7$

$\Rightarrow 2a = 1.3$, i.e., $a = 0.65$

Thus, there are 65% of ^{69}Ga and 35% of ^{71}Ga.

4. Given the first ionization energies, in kJ mol^{-1}, of the elements lithium to neon:

Li	Be	B	C	N	O	F	Ne
519	900	799	1090	1400	1310	1680	2080

(a) Give an equation representing the first ionization energy (IE) of oxygen.

Explanation:

The 1st I.E. of O atom is $O(g) \rightarrow O^+(g) + e^-$.

Do you know?

— I. E. energy measures the strength of the net electrostatic attractive force the nucleus have on the electron after taking into consideration the inter-electronic repulsion exerted by other electrons.

— The species undergoing ionization must be in the gaseous state so that no additional amount of energy is needed to overcome other form of electrostatic attractive forces.

— The 1st I.E. generally increases across the period because of an increase in ENC. Other factors that may lead to a lower-than-expected value for the 1st I.E. includes:

(i) *p* electron has a higher energy level than *s* electron: This factor is used to account for the lower 1st I.E. of aluminum as compared to magnesium.

(ii) Inter-electronic repulsion between paired electrons in a *p* orbital: This factor is used to account for the lower 1st I.E. of oxygen as compared to nitrogen.

— The 1st I.E. decreases down a group because the valence electron that is to be removed resides in a higher-energy principal quantum shell which is farther away from the nucleus.

(b) (i) Explain why the 1st I.E. generally increases across the period.

Explanation:

Across the period, nuclear charge increases but the shielding effect provides by the inner core electrons is relatively constant since the number of inner core electrons is the same. The ENC thus increases. This causes the valence electrons to be more strongly attracted. Hence, 1st I.E. generally increase across the period.

Do you know?

— There is also inter-electronic repulsion between the electrons that reside in the same principal quantum shell, which means there is in fact shielding effect provides by electrons within the same valence principal quantum shell. This shielding effect does affect I.E., but we normally ignore it because some other factors can still be used reasonably well to account for the difference in I. E. values.

(ii) Explain the irregularities in the trend of the 1st I.E. for B and O.

Explanation:

Electronic configuration: B is $1s^22s^22p^1$
O is $1s^22s^22p^4$

Oxygen atom has a higher ENC than boron atom which would cause the 1st I.E. of oxygen to be higher than boron. At the same time, the electron that is to be removed from oxygen atom comes from paired electrons in a *p* orbital. The inter-electronic repulsion between the paired electrons would lower the 1st I.E. of oxygen atom. Now, since the data show that oxygen atom in fact has a higher 1st I.E. as compared to boron atom, this indicates that the predominant factor to account for it is because of the higher ENC of oxygen atom.

Do you know?

— To explain the difference in I.E., it is important to write down the electronic configuration of the species first. Then, inspect the electronic configuration and see which appropriate factor can be used to explain the difference.

— Usually, there is more than one factor affecting the I.E., as shown by the example above. These factors are opposing factors. The predominant factor is the one that supports the experimental data.

(c) An element **W** has successive ionization energies as follows:

786; 1580; 3230; 4360; 16000; 20000; 23600; 29100 kJ mol^{-1}

(i) Provide with explanation which group in the periodic table element **W** belongs to.

Explanation:

The sudden jump in the I.E. values between the 4th and 5th I.E. indicates that the fifth electron that is being removed comes from a lower-energy principal quantum shell. This lower-energy principal quantum shell is nearer to the nucleus and hence the electron is more strongly attracted by the nucleus. The first four electrons that were being removed must be from the valence principal quantum shell. Hence, element **W** is from Group 14.

Why can't we simply use the reason "5th I.E. is higher than 4th I.E. because as more electrons are being removed, the number of electrons decreases but the nuclear charge remains the same. Hence, the inter-electronic repulsion decreases. This thus results in a stronger electrostatic attractive force acting on the remaining electrons, which hence needs more energy to be removed" to account for the difference between the 4th and 5th I.E.?

A: All successive I.E. are increasing values, as can be seen from the given data. The increase is due to the reason that you have provided. But the reason that you have given is not good enough to account for the "significant change" in the I.E. trend. Hence, we have to look for other explanations that can be used to account for our observation.

(ii) Give the valence electronic configuration of an atom of **W**.

Explanation:

Since we do not know which period element **W** comes from, we won't know its valence principal quantum shell number. So generically, the valence electronic configuration of **W** is ns^2np^2.

Do you know?

— The period number represents the highest principal quantum shell or the valence principal quantum shell of the element, while the group number represents the number of valence electrons the element has for Groups 1 and 2. As for Groups 13 to 18, one needs to take the Group Number minus 10 in order to get the number of valence electrons. These do not apply for the *d*-block element.

— *s*-block elements have a valence electronic configuration of ns^1 or ns^2.

— *p*-block elements have a valence electronic configuration of ns^2np^x where x ranges from 1 to 5.

— Noble gas elements have a valence electronic configuration of ns^2np^6.

(iii) Suggest formulas for TWO chlorides of **W**.

Explanation:

Since **W** is from Group 14, and based on the I.E. values, **W** can't be carbon as carbon only has a total of six electrons but there are 8 I.E. values.

W cannot be silicon either because Si only has $SiCl_4$ as a common chloride. So, if **W** is lead, then the two possible chlorides would be $PbCl_2$ and $PbCl_4$.

Do you know?

— Even though lead is a metal and chlorine is a non-metal, $PbCl_4$ is not an ionic compound. It is a simple discrete covalent compound. Why? To form ionic $PbCl_4$, you need to remove four electrons to form a Pb^{4+}. First, it is energetically demanding to remove four electrons. Second, the Pb^{4+} that is formed has very high charge density ($\propto \frac{q_+}{r_+}$) and hence high polarizing power. It would distort the electron cloud of the Cl^- to much extent. This would result in the formation of covalent $PbCl_4$ instead. So, the statement "Metal reacts with non-metal to give ionic compound" is not always true. Another good example to quote is $AlCl_3$.

5. The ionization energies of some successive elements in the periodic table are shown below:

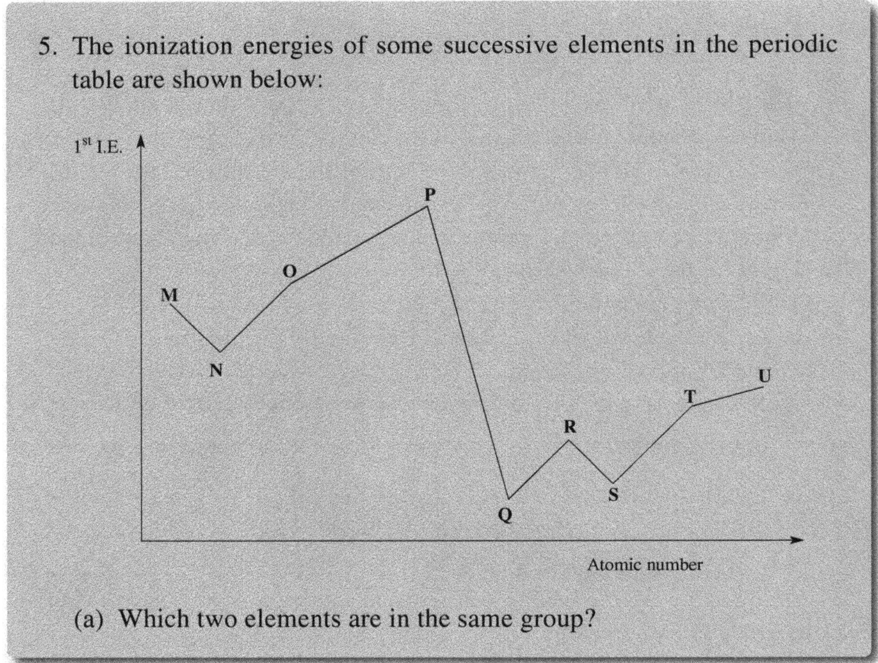

(a) Which two elements are in the same group?

Explanation:

From the graph, element **Q** must come from a different period when compared to element **P**. This is because the drastic decrease in 1st I.E. from **P** to **Q** indicates that the electron that is being removed from **Q** must come from a higher-energy principal quantum shell, which is farther away from the nucleus. Once we have established this relationship, we can proceed to assign the group number for each of the elements:

M is Group 15; **N** is Group 16; **O** is Group 17; **P** is Group 18; **Q** is Group 1; **R** is Group 2; **S** is Group 13; **T** is Group 14; and **U** is Group 15.

So, elements **M** and **U** are from the same group.

Do you know?

— To answer such questions, the most important thing is to find an "anchor element." For example, for the above question, look for a drastic change in the I.E. trend; this would indicate the change of period number.

— Another possibility is the I.E. trend across a period follows what we call the 2-3-3 trend. Meaning? If you take a closer look at the 1st I.E. trend for Period 2 or Period 3 elements, there is a drop in the I.E. at both Group 13 and Group 16.

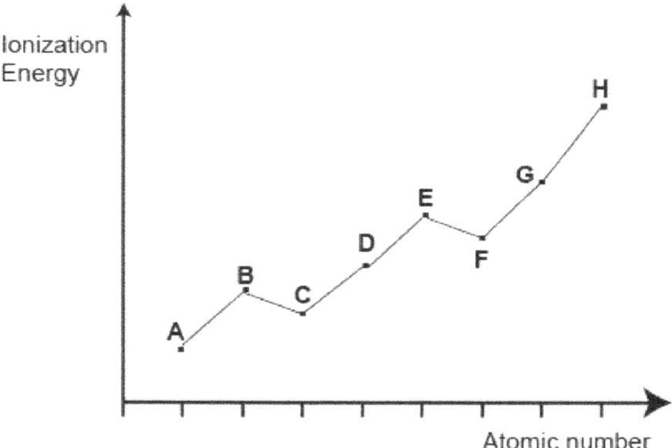

The following table shows the possible species for the above 2-3-3 trend:

	A	B	C	D	E	F	G	H
1st I.E.	Li	Be	B	C	N	O	F	Ne
1st I.E.	Na	Mg	Al	Si	P	S	Cl	Ar
2nd I.E.	Be^{+}	B^{+}	C^{+}	N^{+}	O^{+}	F^{+}	Ne^{+}	Na^{+}
3rd I.E.	B^{2+}	C^{2+}	N^{2+}	O^{2+}	F^{2+}	Ne^{2+}	Na^{2+}	Mg^{2+}
4th I.E.	C^{3+}	N^{3+}	O^{3+}	F^{3+}	Ne^{3+}	Na^{3+}	Mg^{3+}	Al^{3+}
5 th I.E.	N^{4+}	O^{4+}	F^{4+}	Ne^{4+}	Na^{4+}	Mg^{4+}	Al^{4+}	Si^{4+}

Did you notice that the 2-3-3 trend is displaced one atomic number to the right from 1st I.E. to the 2nd I.E.? Then, another one atomic number to the right again from 2nd to 3rd I.E.?

(b) Which element has the largest atomic radius?

Explanation:

Across a period, both I.E. and atomic radius are affected fundamentally by ENC. While down a group, I.E. decreases and atomic radius increases because the valence electrons now sit at a higher-energy principal quantum shell which is farther away from the nucleus.

So, based on the above, element **Q** would have the largest atomic radius because it is the 1st element in the next period after element **P**.

If the atomic radius decreases across a period because of increasing ENC, then why is the atomic radius of Cl 0.099 nm while Ar is 0.192 m, from the data booklet?

A: Chlorine exists as diatomic Cl_2, while argon exists in the monoatomic form. Thus, when we freeze the substance to measure their atomic radius in the solid state, the so-called atomic radius for Cl is actually called the covalent radius. This covalent radius is smaller than the theoretical atomic radius. Why? Because covalent bond pulls two atoms closer together, hence the inter-nuclei separation is shorter. But for argon, the measured atomic radius is known as the van der Waals' radius. Why? van der Waals' forces of attraction pull the argon atoms together in the solid state. But because the van der Waals' forces of attraction are not very strong, the measured atomic radius is much bigger than the theoretical atomic radius of argon.

(c) Explain why the ionization energy of

(i) **N** is smaller than that of **M**;

Explanation:

Valence shell electronic configuration: **M** is ns^2np^3 **N** is ns^2np^4

The electron that is to be removed from **N** atom comes from an electron pair in a *p* orbital. The inter-electronic repulsion between the paired electrons would lower the 1st I.E. of **N** atom as compared to **M**.

(Note: Although **N** has a higher ENC than **M**, the ENC factor "loses" to the inter-electronic repulsion factor.)

(ii) **O** is greater than that of **N**;

Explanation:

Valence shell electronic configuration: **N** is ns^2np^4 **O** is ns^2np^5

Element **O** has a greater ENC than **N**, hence the valence electrons in **O** are more strongly attracted.

(Note: Both elements involve the removal of an electron from an electron pair in a *p* orbital.)

(iii) **P** is greater than that of **O**;

Explanation:

Valence shell electronic configuration: **O** is ns^2np^5 **P** is ns^2np^6

Element **P** has a greater ENC than **O**, hence the valence electrons in **P** are more strongly attracted. In addition, the removal of an electron from **P** would disrupt the symmetrical distribution of the electron density round the nucleus.

(Note: Both elements involve the removal of an electron from an electron pair in a *p* orbital.)

(iv) **P** is greater than that of **N**;

Explanation:

Valence shell electronic configuration: **N** is ns^2np^4 **P** is ns^2np^6

Element **P** has a greater ENC than **N**, hence the valence electrons in **P** are more strongly attracted. In addition, the removal of an electron from **P** would disrupt the symmetrical distribution of the electron density round the nucleus.

(Note: Both elements involve the removal of an electron from an electron pair in a *p* orbital.)

(v) **P** is greater than that of **Q**;

Explanation:

Valence shell electronic configuration: **P** is ns^2np^6 **Q** is $(n + 1)s^1$

The electron that is removed from element **Q** comes from a higher-energy principal quantum shell, which is farther away from the nucleus. Hence, the valence electron in **Q** is less strongly attracted by the nucleus.

(vi) **R** is greater than that of **Q**;

Explanation:

Valence shell electronic configuration: **Q** is $(n + 1)s^1$ **R** is $(n + 1)s^2$

Element **R** has a greater ENC than **Q**, hence the valence electrons in **R** are more strongly attracted.

Q The electron that is removed in element **R** comes from an electron pair in the *s* orbital. So, why didn't the inter-electronic repulsion factor help to lower the I.E.?

A: The inter-electronic repulsion between the pair of electrons in the *s* orbital is not significant enough to lower the I.E. here, but is significant enough to lower the I.E. if the pair of electrons resides in a *p* orbital or even a *d* orbital for a transition element. This is because a *p* orbital is only about 33% the capacity of an *s* orbital. Hence, the inter-electronic repulsion is much more prominent in a smaller space. This is, similar for a *d* orbital, which has only a capacity of about 20% the size of an *s* orbital.

(vii) **T** is smaller than that of **U**.

Explanation:

Valence shell electronic configuration: **T** is $(n + 1)s^2(n + 1)p^2$
U is $(n + 1)s^2(n + 1)p^3$

Element **U** has a greater ENC than **T**, hence the valence electrons in **U** are more strongly attracted.

CHAPTER 2

CHEMICAL BONDING

(a) Solids with crystalline structures may be classified as *molecular*, *giant covalent*, or *ionic*. With reference to each crystalline structure, explain how *low melting point, high melting point, and non-conduction of electricity* are related to bonding and structure.

Explanation:

Molecular compound consists of simple discrete molecules attracted to each other via intermolecular forces. Each of the molecules consists of atoms bonded together by covalent bonds, which is difficult to break under normal conditions. When a melting or boiling process is carried out, the heat energy is used to overcome the weak intermolecular forces, which can be instantaneous dipole–induced dipole (for non-polar molecules), permanent dipole–permanent dipole (for polar molecules), or hydrogen bonding (for molecules that are capable of forming this type of intermolecular force). These account for the low melting or boiling point. In addition, because the molecules are overall uncharged particles, these types of compounds are non-conductors of electricity.

Giant covalent compound consists of atoms bonded together by covalent bonds which are difficult to break under normal conditions. Hence, when the melting or boiling process is carried out, the heat energy is use to overcome the strong covalent bond. This accounts for the high melting or boiling point. The giant covalent lattice structure consists of atoms rigidly bonded, hence these electrically neutral atoms cannot move as charge carriers. Therefore, most of the giant covalent compounds are non-conductors of electricity. There are some giant covalent compounds, such as graphite, which are capable of conducting electricity. This is because of the presence of delocalized electrons that can move under the influence of an electric field.

Ionic compound consists of charged cations and anions bonded together by ionic bonds, which are difficult to break under normal conditions. Hence, when the melting or boiling process is carried out, the heat energy is use to overcome the strong ionic bond. This accounts for the high melting or boiling point. The giant ionic lattice structure consists of ions rigidly bonded, hence these charged particles cannot move when in the solid state. But when melted into the molten state, these ions can act as charge carriers. Hence, this accounts for the electrical conductivity of molten ionic compound.

Do you know?

For simple molecular compound:

— A lot of students think that when we melt or boil simple molecular compounds, we are actually breaking covalent bonds. THIS IS INCORRECT!

— The strength of instantaneous dipole–induced dipole interaction depends on:

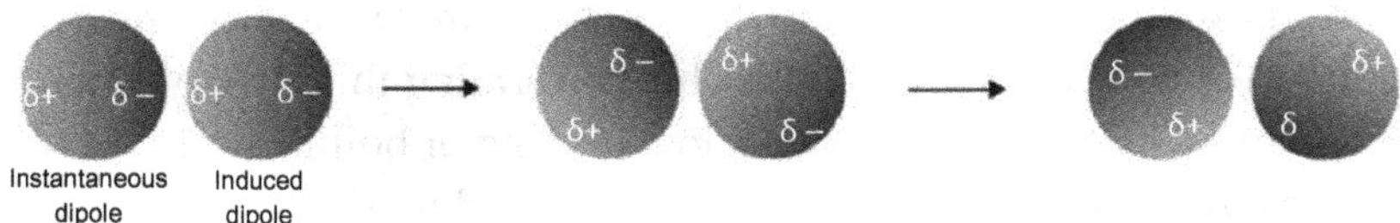

(i) *The number of electrons the molecule has.* The greater the number of electrons, the more polarizable the electron cloud, hence the more extensive is the id–id interaction. For example, I_2 is a solid at room temperature, Br_2 is a liquid, and Cl_2 is a gas.

(ii) *The surface area of contact between the molecules.* If two different molecules have the same number of electrons, but one is more

(*Continued*)

(Continued)

linear than the other, the id–id interaction between the linear molecules is more extensive than the spherical one. For example, pentane has a higher boiling point than dimethylpropane.

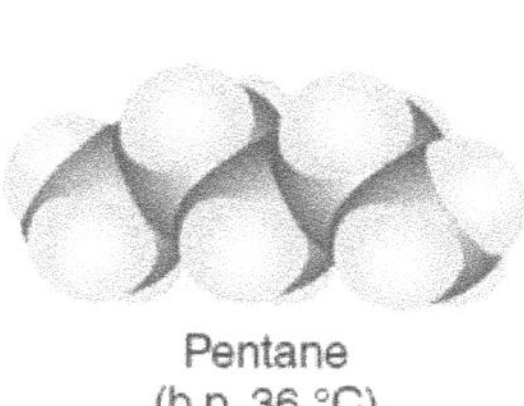

Pentane
(b.p. 36 °C)

Dimethylpropane
(b.p. 10 °C)

— The strength of permanent dipole–permanent dipole interaction depends on the polarity of the molecule, which is measured by the net dipole moment of the molecule. The net dipole moment of a molecule is dependent on:

(i) *The polarity of each of the bonds in the molecule*, and the polarity of a bond is dependent on the difference in the electronegativity between the two bonding atoms. The more polar the bond, the greater the dipole moment of the bond. In addition, dipole moment is a vector quantity.

(ii) *The molecular shape of the molecule.* A molecule can have polar bonds but if the molecular shape is symmetrical, the various dipole moments of the bonds may cancel out each other. For example, CCl_4 is a non-polar molecule although C-Cl bond is polar because CCl_4 has a tetrahedral shape.

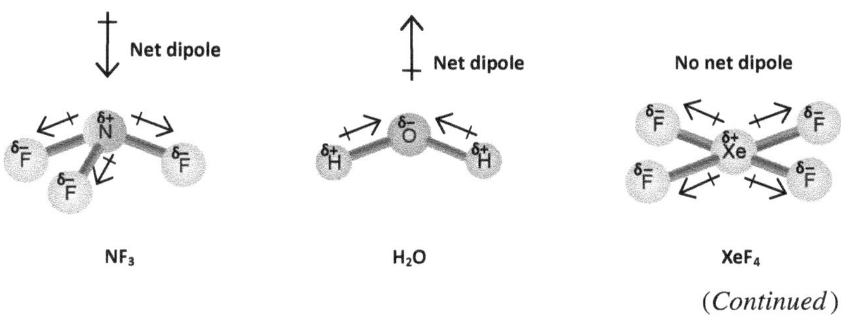

(Continued)

(*Continued*)

— The strength of hydrogen bond depends on:

(i) *The type of hydrogen bond that is being formed.* Hydrogen bond that is formed between H-F:----H-F is stronger than that between H-O:----H-O, which in turn is stronger than H-N:----H-N. The reason is because of an increase in the polarity of the bond: H-N< H-O< H-F.

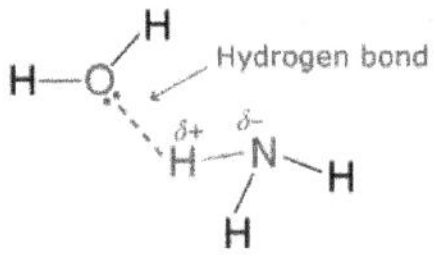

(ii) *The number of hydrogen bonds that the molecule can form.* For example, H_2O (100°C) molecule has higher boiling point than HF (~19°C) because a H_2O molecule can form an average of two hydrogen bonds, whereas HF can only form one hydrogen bond. Another example is CH_3OH, which has a boiling point of ~65°C as compared to H_2O. This is because CH_3OH can, on the average, only form one hydrogen bond per molecule.

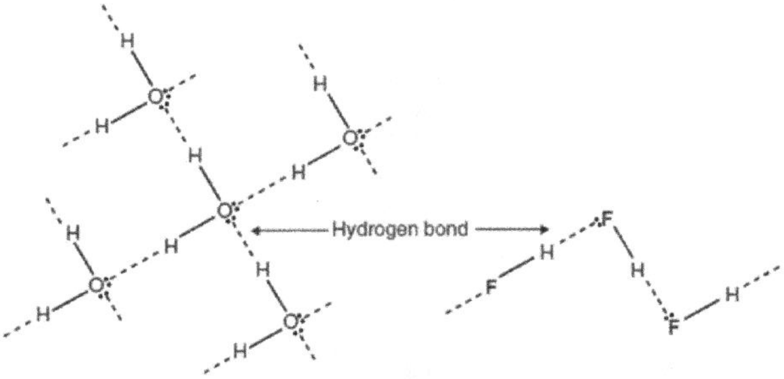

Q How does the sharing of electrons hold two atoms together?

A: Both nuclei would attract the oppositely charged shared electrons. At the same time, there would also be inter-electronic repulsion between the shared electrons, in addition to the inter-nuclei repulsion. If the attractive force can overpower the repulsive forces, then there would be a net attractive force to pull the two atoms together.

Q Does the non-bonding electrons on each atom affect the strength of the covalent bond formed?

A: Yes, of course. Take for instance, in F-F molecule, the inter-electronic repulsion between the lone pair of electrons from both F atoms is so prominent that it actually weakenes the F-F bond. This is exemplified in a smaller bond energy for the F-F bond (158 kJ mol^{-1}) as compared to Cl-Cl which has a bond energy of 244 kJ mol^{-1}. So, the whole idea here is that if we use the "effective overlap of orbitals" concept to try to account for the strength of the F-F bond, we would find that this concept should explain that the F-F bond is stronger and hence has a more endothermic bond energy. But the data speak otherwise. Thus, there must be other contributing factors that lead to this observation. In a nutshell, what we are trying to advocate here is that, it is very important to understand fundamentally at the atomic level, what actually are the various factors that are interplaying. The fundamental concepts are, electrons are negatively charged while protons are positively charged and they attract each other. But like charges also repel each other. An atom doesn't know anything about orbitals; it doesn't even know whether it should form an ionic or covalent compound! All these concepts are invented by us, human beings!

Do you know?

For giant covalent compound:

— The strength of covalent bond depends on:

(i) *A sigma bond being stronger than a pi bond.* This is because a sigma bond is formed via head-on overlap of orbitals and as a result, there is accumulation of shared electron density

(*Continued*)

(*Continued*)

within the inter-nuclei region whereas for a pi bond, the shared electron density is out of the inter-nuclei region, as it is a result of side-on overlap of orbitals. With shared electron density being accumulated within the inter-nuclei region, there is minimal inter-nuclei repulsion.

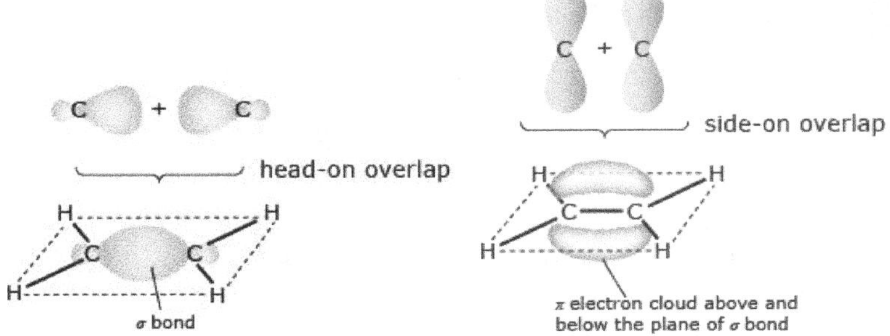

(ii) *The number of covalent bonds forms between the two atoms*. For example, C=C is stronger than C–C because there are more shared electrons being attracted by the two positively charged nuclei.

(iii) *The effective overlap of atomic orbitals between the two atoms*. Covalent bond can be perceived as the overlapping of atomic orbitals. If an atom uses a bigger atomic orbital for covalent bond formation than another atom, then the overlap of the orbitals is less effective. This arises because a bigger atomic orbital is more diffuse. For example, the H-Cl bond is stronger than the H-Br bond because the Br atom uses a bigger orbital, which is more diffuse, to form a bond with the hydrogen atom. Other examples include comparing C-Cl and C-Br, where C-Br has a weaker bond than C-Cl.

(iv) *The polarity of the bond*. The more polar a bond, the stronger the covalent bond. For example, the H-O bond is stronger than the H-N bond because the greater dipole that is created in the H-O bond causes an additional stronger attractive force between the dipoles.

Do you know?

For ionic compound:

— The strength of ionic bond depends on:

(i) *The charges of the cation and anion.* For example, the ionic bond in MgO is stronger than Na_2O because Mg^{2+} is doubly positively charged. Or, the ionic bond in MgO is stronger than that in MgF_2 because O^{2-} is doubly negatively charged.

(ii) *The sizes of the cation and anion.* For example, the ionic bond in MgO is stronger than that in CaO because Mg^{2+} has a smaller cationic radius than Ca^{2+}, therefore it has a greater charge density ($\propto \frac{q_+}{r_+}$). Or, the ionic bond in MgO is stronger than that in MgS because O^{2-} has a smaller anionic radius than S^{2-}.

Q How does a cation actually attract an anion?

A: A cation does not actually carry a "+" charge and attract the "–" charge of an anion. When a cation is near an anion, the nucleus of the cation would attract the electron cloud of the anion (opposites attract), and vice versa. But at the same time, there would also be repulsive forces between both particles' electron clouds and their nuclei. Hence, an ionic bond is actually the net electrostatic attractive force that is formed. If the repulsive forces overpower the attractive force, then the two oppositely charged particles would not be able to form the ionic compound. So, in a nutshell, a chemical bond is actually the result of the formation of a net electrostatic attractive force.

(b) Explain why the boiling points of the noble gases increase from helium to xenon.

Explanation:

The noble gases consist of monoatomic particles. Hence, they are non-polar. The increase of boiling points from helium to xenon is because of an increase in the strength of instantaneous dipole–induced dipole interaction. This is caused by an increase in the number of electrons from helium to xenon, which makes the electron cloud more polarisable.

2. (a) Boron trihydride, BH_3, reacts with trimethylamine, $(CH_3)_3N$, to form a 1:1 adduct.
   (i) Using BH_3 and $(CH_3)_3N$ as examples, explain how the shapes of simple molecules can be determined by simply considering the repulsions between electron pairs.

Explanation:

The following are the dot-and-cross diagrams of BH_3 and $(CH_3)_3N$:

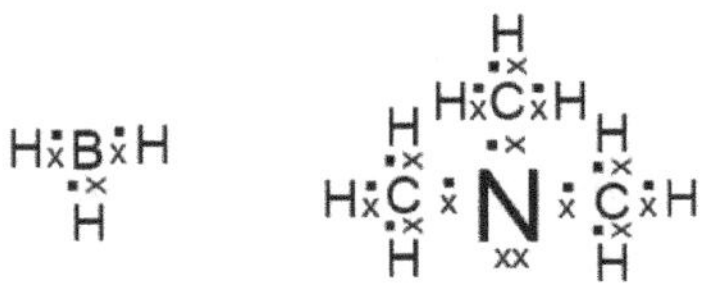

```
                         H
                         •x
                      H H:C:H H
 H:B:H                •x  •x  •x
  •x                H:C : N : C:H
  H                   •x  xx  •x
                      H       H
```

Consider BH_3:

No. of regions of electron densities = 3

According to the Valence Shell Electron Pair Repulsion (VSEPR) theory, electron pairs would spread out as far apart as possible to minimize inter-electronic repulsion; the electron pair geometry (EPG) is trigonal planar. Since there is no other lone pair of electrons, the molecular geometry (MG) is also trigonal planar.

Consider the carbon atom in $(CH_3)_3N$:

No. of regions of electron densities = 4

According to the VSEPR theory, electron pairs would spread out as far apart as possible to minimize inter-electronic repulsion; the EPG is tetrahedral.

Since there is no other lone pair of electrons, the MG is also tetrahedral.

Consider the nitrogen atom in $(CH_3)_3N$:

No. of regions of electron densities = 4

According to the VSEPR theory, electron pairs would spread out as far apart as possible to minimize inter-electronic repulsion; the EPG is tetrahedral.

Since there is one lone pair of electrons, the MG is trigonal pyramidal.

Do you know?

— The VSEPR theory is a simple and easy-to-use theory as it simply makes use of the fact that electrons are negatively charged particles, so when you have electron pairs present in the same region of space, it is natural these electron pairs would have to spread out so as to minimize inter-electronic repulsion.

— The MG or shape of a molecule is only dependent on the relative orientation of the various nuclei around the central atom. It is independent of the lone pair of electrons around the central atom. This is because when we determine the shape of a molecule experimentally, it is the relative positioning of the nuclei in space that is revealed to us. We would not be able to locate the electrons as they are simply too small to be "seen" as compared to the more massive nucleus!

So does it mean that the lone pair of electrons around the central atom is not important in affecting the overall shape of the molecule?

A: Of course not! The shape of the molecule can only be determined after obtaining the EPG. What is important to take note here is that the MG of a molecule may differ from its EPG! So have you remembered the following table?

Number of Electron Pairs	Electron–Pair Geometry	No. of Lone Pairs	Molecular Geometry	Examples
2	Linear	0	Linear (180°)	Cl—Be—Cl, O=C=O
3	Trigonal planar	0	Trigonal planar (120°)	BF_3
		1	Bent	SO_2
4	Tetrahedral	0	Tetrahedral (109.5°)	CH_4
		1	Trigonal pyramidal	NH_3
		2	Bent	H_2O
5	Trigonal bipyramidal	0	Trigonal bipyramidal (120°)	PCl_5
		1	Distorted tetrahedral (or seesaw or sawhorse)	SF_4
		2	T-shaped	ClF_3
		3	Linear	$[IF_2]^-$
6	Octahedral	0	Octahedral (90°)	SF_6
		1	Square pyramidal	BrF_5
		2	Square planar	XeF_4

(ii) Name the type of bond that is formed between BF_3 and $(CH_3)_3N$ in the adduct.

Explanation:

The following is the adduct formed between BF_3 and $(CH_3)_3N$:

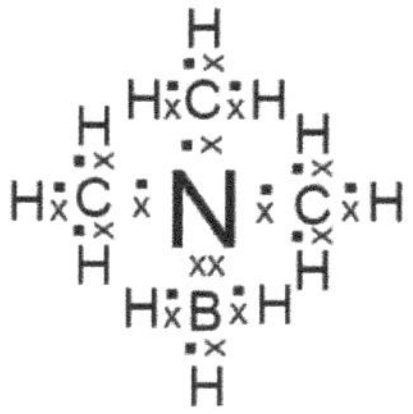

As can be seen from the dot-and-cross diagram, the pair of electrons from the N atom to the B atom belongs solely to the N atom. Hence, the type of bond that is formed between BF_3 and $(CH_3)_3N$ is a dative covalent bond.

Do you know?

— An adduct is a single reaction product formed from a direct addition of two or more distinct molecules, containing all atoms from the component molecules. It is not a dimer! In a dimer, the two molecules that create the dimer is exactly of the same type, but not in an adduct.

So, is a dative covalent bond of the same strength as a covalent bond formed between the same two atoms?

A: No! This is because the donor atom of the dative covalent bond is not 100% "altruistic" to "sacrifice" two electrons for sharing. Hence, the so-called shared electrons would be more attracted to the donor's atom.

(b) Boron and carbon are adjacent to each other in the periodic table. There are some B–N compounds known to be *isoelectronic* with C–C compounds. An example would be boron nitride, an electrical insulator, which has a planar hexagonal layered structure of alternating boron and nitrogen atoms, similar to graphite.

(i) Explain the meaning of the term *isoelectronic*.

Explanation:

Two species are isoelectronic if they have the same number of electrons.

Do you know?

— If two species are isoelectronic, it does not mean that they have the same electronic configuration. For example, Fe^{2+} (24 electrons) and Cr (24 electrons) are isoelectronic but their electronic configurations are not the same.

— Another possible mistake that students normally would make is to use the term *isoelectric* here. Isoelectric refers to the pH point whereby the species is a zwitterion, i.e., an electrically neutral particle containing an equal number of positive and negative charges.

What is the advantage of knowing that two or more species are isoelectronic?

A: If two species are isoelectronic, then the amount of inter-electronic repulsion is the same since both species contain the same number of electrons. Then, if these two species have different number of protons, the one that has a higher proton number would have its electron cloud more strongly attracted. Therefore, we can infer the ionisation energy (I.E.) value or size of the species here.

(ii) Give the type of bonding which is present *within* the layers.

Explanation:

Since the structure of boron nitride is similar to that of graphite, the B and N atoms must be held together by B–N covalent bonds.

Do you know?

— The structure of boron nitride is similar to that of graphite. Since the carbon atom in graphite is trigonal planar in shape, it should be sp^2 hybridized with an unused *p* orbital containing a lone electron. Similarly, the B and N atoms in boron nitride are also sp^2 hybridized, as we were told that it "has a planar hexagonal layered structure of alternating boron and nitrogen atoms." But different from graphite, the N atom has an unused *p* orbital containing a lone pair of electrons whereas the *p* orbital of the B atom is empty.

— Since every atom in boron nitride has a *p* orbital parallel to one another, these *p* orbitals can overlap sideways. Hence, the electrons in the *p* orbital can move under the influence of an electric field that is applied along the plane. If the electric field is applied perpendicular to the plane, electric current would not flow because there are gaps between the planes.

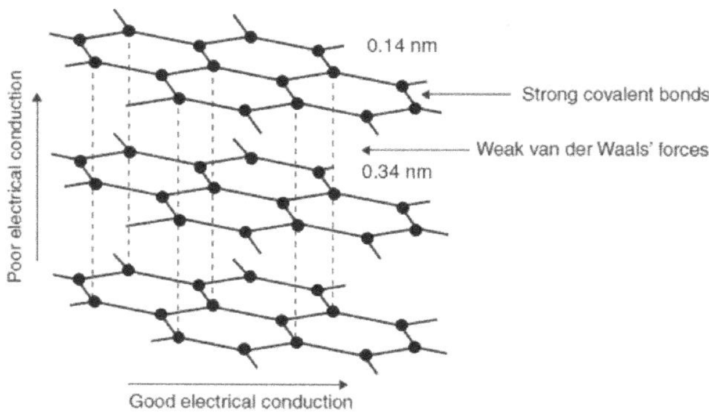

Q What is hybridization?

A: Hybridization model was formulated mathematically to account for observed experimental data (on bond angle, length, and bond strength). Apart from atomic orbitals, hybrid orbitals are also used to form covalent bonds. A specific number and type of atomic orbitals combine to give exactly the same number of hybrid orbitals. There are three common types of hybridization:

(i) sp^3 hybridization

Central atom forms four sigma bonds or there are four regions of electron densities. Take for example, the C atom:

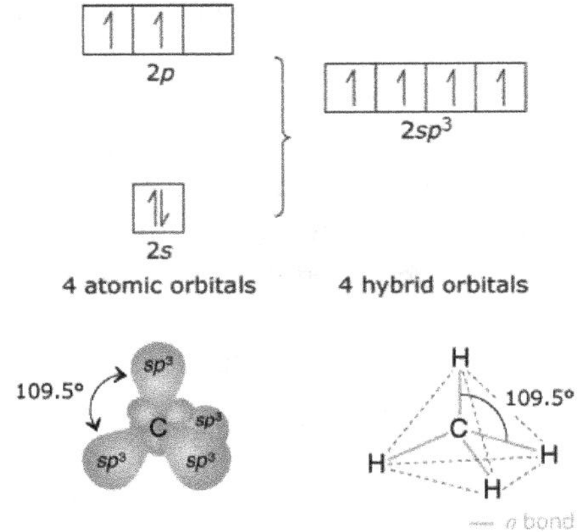

A sp^3 hybridized atom has bond angles close to 109.5°. This would mean that if we know that the central atom has bond angles close to 109.5°, the best postulate is that it is a sp^3 hybridized atom! Or if we see that the central atom has a tetrahedral EPG, it is likely to be a sp^3 hybridized atom.

(ii) sp^2 hybridization

Central atom forms three sigma and one pi bonds. Take for example, the C atom:

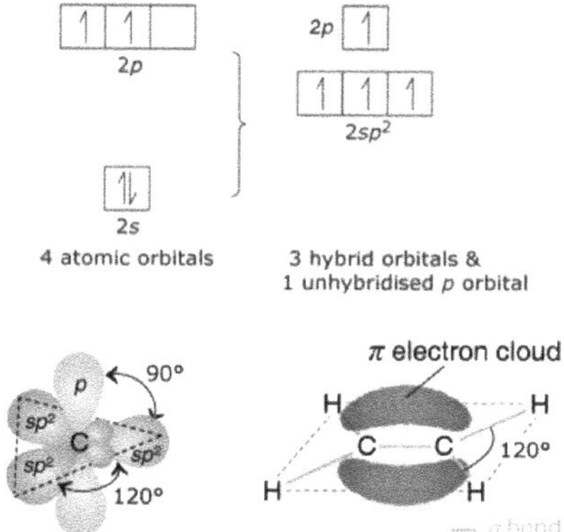

A sp^2 hybridized atom has bond angles close to 120°. This would mean that if we know that the central atom has bond angles close to 120°, the best postulate is that it is a sp^2 hybridized atom! Or, if we see that the central atom forms one double bond and two single bonds, it is also likely to be a sp^2 hybridized atom. Or, if we see that the central atom has a trigonal planar EPG, it is likely to be a sp^2 hybridized atom.

(iii) sp hybridization

Central atom forms two sigma and two pi bonds. Take for example, the C atom:

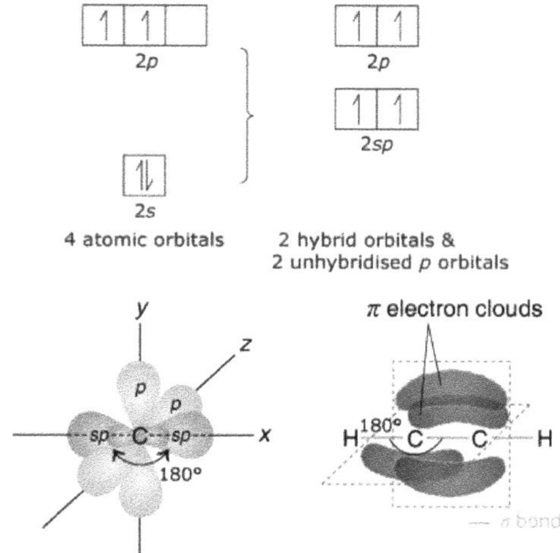

A *sp* hybridized atom has an angle close to 180°. This would mean that if we know that the central atom has an angle close to 180°, the best postulate is it is a *sp* hybridized atom! Or, if we see that the central atom has a linear EPG, it is likely to be a *sp* hybridized atom.

(iii) Give the type of interaction *between* the layers.

Explanation:

From the outset, a layer of atoms is none other than a mass of electron cloud. Since there is no covalent bond between the layers, what hold the different layers together would be intermolecular forces of the instantaneous dipole–induced dipole type.

Do you know?

— Since the layers are being held together by intermolecular forces of the instantaneous dipole–induced dipole type, which is not strong in nature, the different layers can slide over each other easily. This allows boron nitride to be used as a lubricant, like graphite.

(iv) Give a possible use in which this compound could replace the industrial usage of the corresponding carbon compound.

Explanation:

Boron nitride can replace graphite as a high-temperature lubricant.

Do you know?

— Since the atoms in boron nitride are bonded by a strong B-N covalent bond, which will only break at high temperature, it can be used as a lubricant at high temperature.

(v) When heated under high pressure, this form of boron nitride is converted into another form which is an extremely hard solid. Suggest the type of structure adopted by this new material.

Explanation:

In this new material, each of the boron and nitrogen would adopt a tetrahedral shape. Among the four covalent bonds that an N atom forms with four B atoms, one of the N→B bond is a dative covalent bond. The overall structure of this new material is very similar to the structure of diamond.

If we have not come across the chemistry of boron nitride before, how would we know how to answer this question?

A: In this chapter, you need to know the structures and bonding of some basic substances such as diamond, graphite, sodium chloride, water, etc. In the very beginning, the question has already told you that boron nitride is similar to graphite. So, you need to use the physical and chemical properties of graphite here. Now, since diamond is related to graphite in terms of allotropy, we thus can infer that the planar layered structure of boron nitride can be transformed into the tetrahedral form. Well, this is applying what we have learned into new context.

3. Sulfur reacts with nitrogen to yield nitrogen disulfide (NS_2), an electronically symmetrical gas phase radical which is highly unstable. This radical undergoes a further reaction to give a cyclic molecule, dinitrogen disulfide (N_2S_2). The N_2S_2 is explosive in nature, and can also be heated to give the polymeric (SN)x, which is metallic in nature. At very low temperatures (0.33 K), the polymeric compound is a superconductor.
   (a) With the aid of a dot-and-cross diagram, deduce the shape of NS_2 and N_2S_2.

Explanation:

The possible dot-and-cross diagrams of NS_2 and N_2S_2:

Consider the N atom in NS_2:

No. of regions of electron densities = 3

According to the VSEPR theory, electron pairs would spread out as far apart as possible to minimize inter-electronic repulsion; the EPG is trigonal planar.

Since there is one lone electron, which is also considered a region of electron density, the MG is bent or V-shape.

Consider the N atom in N_2S_2:

No. of regions of electron densities = 3

According to the VSEPR theory, electron pairs would spread out as far apart as possible to minimize inter-electronic repulsion; the EPG is trigonal planar.

Since there is one lone pair of electrons, the MG is bent or V-shape.

Consider the S atom in N_2S_2:

No. of regions of electron densities = 4

According to the VSEPR theory, electron pairs would spread out as far apart as possible to minimize inter-electronic repulsion; the EPG is tetrahedral.

Since there are two lone pair of electrons (actually it consists of one lone pair and one lone electron), the MG is bent or V-shape.

Do you know?

— The dot-and-cross diagram of NS_2 is very similar to that of NO_2. This is quite logical since both O and S belong to the same group.

— When we draw the dot-and-cross diagram, the basic guideline is to let as many atoms in the structure achieve an octet configuration as possible. If we can't satisfy everyone, then let it be. If we need to expand beyond the octet configuration, then take note that only elements that come from Period 3 and below can do it. This is because these elements possess vacant low-lying orbitals that can be readily used for "extra" covalent bond formation.

Q If a lone electron can be considered as a region of electron density, then why don't we consider three lone electrons as three separate regions of electron density? Why must we consider it as two regions of electron density instead?

A: Well, we can pair up electrons and put them into the same region of space rather than separating them and put them into different regions of space. Imagine if we have too many electrons, how much individual space could we give to each of the electrons? In addition, do not forget that when we pair up two electrons, they would have opposite spins and these opposite spins would creates opposite magnetic fields to attract each other. So, it seems that nature knows the concepts of effectiveness and efficiency too well!

Would a lone electron exert the same amount of repulsive force as a lone pair of electrons?

A: Of course not! Hence, the bond angle would be greater for a central atom that has lone electron–bond pair repulsion when compared to another that has lone pair–bond pair repulsion. Take for instance, NO_2 has a trigonal planar EPG, the bond angle of NO_2 is actually greater than 120° because the bond pair-bond pair repulsion is much greater than the lone electron-bond pair repulsion. But for NH_3, which has a tetrahedral EPG, the bond angle is less than 109.5°, about 107°, because the lone pair-bond pair is much greater than its bond pair-bond pair repulsion. As for H_2O, the lone pair-lone pair repulsion is much greater than its lone pair-bond pair repulsion which in turn is greater than its bond pair-bond pair repulsion. This accounts for the bond angle of H_2O to be about 105°

(b) Give a possible explanation why both NS_2 and N_2S_2 are highly unstable.

Explanation:

The N atom in NS_2 does not have an octet configuration and is electron-deficient as it is a radical, possessing an unpaired electron. This makes NS_2 possible to form a bond with another species.

As for N_2S_2, the molecule is highly unstable because a four-membered ring structure contains too much ring strain. If the angles of N–S–N or S–N–S bonds are about 90°, there would be too much inter-electronic repulsion between the various bond pairs of electrons. Hence, the weakened bonds need less energy to break, which accounts for its high reactivity.

When a radical possesses an unpaired electron, why is it highly reactive?

A: If a radical has an unpaired electron, it would mean that the orbital where this unpaired electron sits still has the "capacity" to "welcome" another electron. This would make the radical reactive, isn't it? And because of the "lack of another electron," we can thus say that a radical is an electron-deficient species. And once a radical forms a bond with another radical, energy is released, the energy level of the product is lowered as compared to the reactant. Hence, the product is more stable than the reactant!

(c) Give a possible structure of the polymeric (SN)x, depicting its unidirectional electrical conducting property.

Explanation:

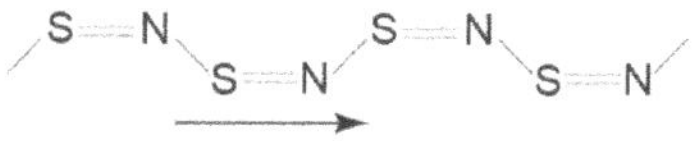

S=N–S=N–S=N–S=N

Electrons flow along the N-S-N bonds unidirectionally.

Electrons can flow along the N–S–N backbone because each of the atoms has a *p* orbital parallel to one another. This allows all the *p* orbitals to overlap sideways to form a continuous pi network. Thus, when an electric field is applied along the N–S–N backbone, electrons can flow through this pi network.

Do you know?

— When you have alternating double bonds, we call this phenomenon 'double bond conjugation.'

(d) Give a possible reason why the polymeric compound becomes super conducting at low temperatures.

Explanation:

At low temperatures, all the atoms have very little amount of kinetic energy; this would mean that atomic vibration is minimal. Hence, the pi electrons would encounter less vibrational resistance to move through the pi network. In addition, because of minimal atomic vibration, the overlapping of the *p* orbitals would be more effective.

(e) Give a possible reason why it is possible for such a polymeric compound form between sulfur and nitrogen.

Explanation:

Both sulfur and nitrogen atoms have similar electronegativity values. This allows bonding electrons to be more equally shared and the non-bonding electrons, such as the pi electrons, to flow freely through the pi network, without specifically being attracted more strongly to a particular atom. If there is specific point of strong attractive force, this would hinder the movement of electrons.

4. This question is concerned with the shapes of molecules and the forces between them.
   (a) Depict with clear diagrams, the shapes of the following molecules:
      (i) CH_4; (ii) H_2O

Explanation:

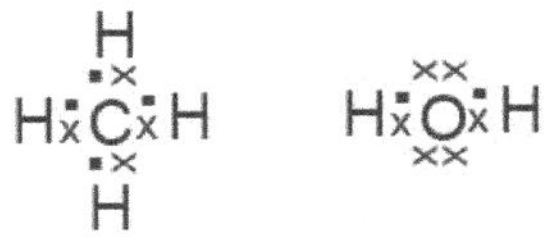

```
    H
   •x             xx
 H x• C •x H    H x• O •x H
   •x             xx
    H
```

Consider the C atom in CH_4:

No. of regions of electron densities = 4

According to the VSEPR theory, electron pairs would spread out as far apart as possible to minimize inter-electronic repulsion; the EPG is tetrahedral.

Since there is no lone pair of electrons, the MG is tetrahedral.

Consider the O atom in H_2O:

No. of regions of electron densities = 4

According to the VSEPR theory, electron pairs would spread out as far apart as possible to minimize inter-electronic repulsion; the EPG is tetrahedral.

Since there are two lone pair of electrons, the MG is bent or V-shape.

The actual way to depict the shape of CH_4 and H_2O is as follows:

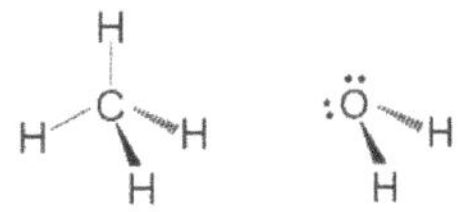

The lines represent bonds sitting on the plane; the solid wedge represents bond going into the plane from the front; and the dotted wedge represents bond going into the plane from behind the plane.

Do you know?

— All CH_4, NH_3, and H_2O have the same EPG (i.e., tetrahedral) but different molecular shape because of the presence of different numbers of lone pair of electrons.

— The bond angle decreases from CH_4 to NH_3 to H_2O because of the following reason: LP–LP repulsion > LP–BP repulsion > BP–BP repulsion. Lone pair electrons (LP) exert greater repulsive force because they are localized on one atom, whereas bond pair electrons (BP) are delocalized between two shared atoms. For the presence of every LP, the bond angle decreases approximately by 2°. This accounts for the bond angle of NH_3 to be 107° and H_2O to be 105°. Regular tetrahedral has a bond angle of 109.5°.

(b) Explain the main features of the theory that is used to predict the shapes of these molecules.

Explanation:

The theory that is used to predict shape is known as the Valence Shell Electron Pair Repulsion theory or VSEPR for short. According to the VSEPR theory, the various electron pairs around a central atom would spread out as far apart as possible to minimize inter-electronic repulsion. The spatial distribution of these electron densities, which can be bond pairs, lone pairs or even lone electron, is known as the electron pair geometry.

Now, since lone pairs of electrons cannot be seen as compared to the massive nuclei present in the molecule, the molecular geometry is predicted without taking into consideration the positions of these lone pair of electrons. Hence, the shape of a molecule is in fact the spatial positioning of the various nuclei.

The lone pair of electrons is important as it affects the bond angle of the molecule. This is because the LP exert greater repulsive force than the bond pair as it is localized on one atom. Whereas BP is delocalized between two shared atoms. As a result of this difference in repulsive force, we have: LP–LP repulsion > LP–BP repulsion > BP–BP repulsion, as a guiding principle when predicting the bond angles in a molecule.

(c) The boiling points and molar masses of some first row hydrides are tabulated below.

Substance	Boiling point/K	Molar mass/g mol^{-1}
CH_4	109	16
NH_3	240	17
H_2O	373	18

(i) Explain the difference in boiling points between NH_3 and CH_4 in terms of the structure and bonding.

Explanation:

Both NH_3 and CH_4 exist as simple discrete molecular compounds held together by weak intermolecular forces. The intermolecular forces between the non-polar CH_4 molecules are of the instantaneous dipole–induced dipole type. This is much weaker than the hydrogen bond between the NH_3 molecules. Hence, CH_4 has a lower boiling point than NH_3.

(ii) Why does H_2O have a higher boiling point than NH_3?

Explanation:

Both NH_3 and H_2O exist as simple discrete molecular compounds held together by hydrogen bonds. The hydrogen bonds between the H_2O molecules are more extensive than those between NH_3. This is because a H_2O molecule can form, on the average, two hydrogen bonds per molecule, whereas a NH_3 molecule can only form one hydrogen bond per molecule. Hence, the boiling point of water is higher than ammonia.

(d) 1,2-dihydoxybenzene (compound A) has a boiling point of 518 K, but its isomer 1,4-dihydroxybenzene (compound B) has a boiling point of 558 K.

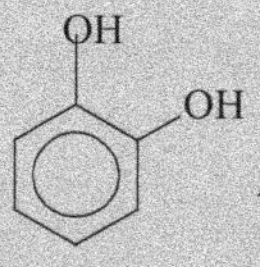

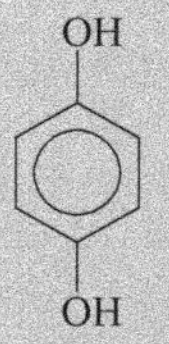

Give one reason why 1,4-dihydroxybenzene has the higher boiling point.

Explanation:

Both molecules contain OH groups capable of forming hydrogen bonds intermolecularly. But the two OH groups of compound **A** are too close to each other. This causes the two OH groups to be able to form intramolecular hydrogen bonds. The formation of intramolecular hydrogen bonds limits sites available for the formation of intermolecular hydrogen bonds, hence this decreases the boiling point of compound A. For compound **B**, the two OH groups are too far away to form intramolecular hydrogen bonds.

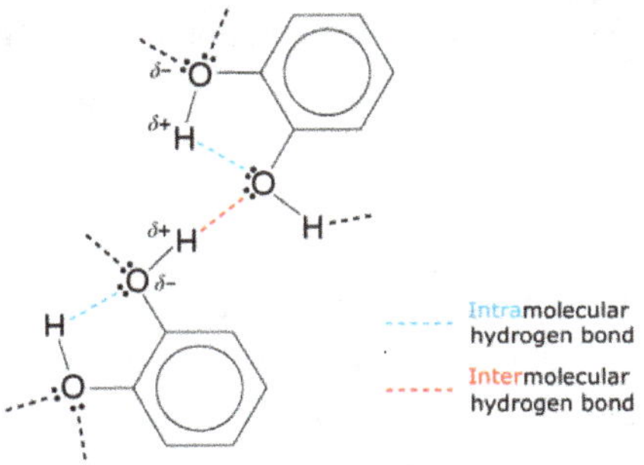

Do you know?

— The formation of a hydrogen bond needs a lone pair of electrons sitting on an O, F, or N atom. And this lone pair of electrons is attracted to a H atom that is covalently bonded to O, F, or N. Thus, if a lone pair of electrons or a H atom is used up to form an intramolecular hydrogen bond, then it cannot be used to form intermolecular hydrogen bond.

5. (a) Molecular crystals composed of simple discrete molecules are held together in a regular array.
   (i) With references to bromine, Br_2, and bromomethane, $CH_3Br(s)$, which are crystalline solids at low temperatures, describe and explain the types of intermolecular forces for holding the molecules together in such crystals.

Explanation:

Br_2 is a non-polar molecule, hence the intermolecular forces that hold the molecules together in the crystal are of the instantaneous dipole–induced dipole type. As for CH_3Br, it is a polar molecule as there is a net dipole moment. Hence, the intermolecular forces are of the permanent dipole–permanent dipole type.

Are there any id-id interaction between the CH_3Br molecules? Is it significant?

A: There is id-id interaction between polar molecules, but it is considered as insignificant here. As by comparing the pd-pd of CH_3Br and the id-id of Br_2, we can be "satisfied" to have reasonably account for the observed experimental

(ii) Give a reason for one physical property associated with molecular crystals.

Explanation:

We would expect the crystal to be soft because of the weak intermolecular forces that hold the molecules together. In addition, the melting point is low as compared to the ionic or giant covalent compound.

(b) In contrast, substances such as silicon dioxide and sodium chloride consist of giant lattice crystal structures, which do not contain simple discrete molecules. For each of these substances:
(i) describe their crystalline structures;

Explanation:

For NaCl, it consists of a giant ionic lattice crystal structure whereby the ions are held together by strong ionic bonds.

As for SiO_2, it consists of a giant covalent lattice crystal structure whereby the atoms are held together by strong covalent bonds formed between Si and O atoms.

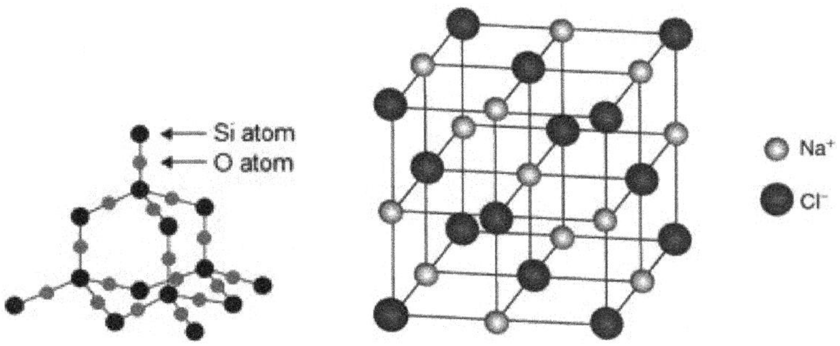

Do you know?

— When you are asked to draw the dot-and-cross diagram of an ionic compound, we need to take note that (i) there should not be any electrons surrounding the elemental symbol of the cation, and (ii) the electrons that are gained by the anion is represented as a different symbol from the original valence electrons that it has. For example,

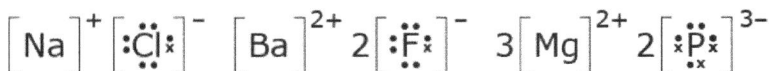

NaCl	BaF_2	Mg_3P_2
$[Na]^{+}[:\ddot{\underset{\cdot\cdot}{Cl}}\times]^{-}$	$[Ba]^{2+}\,2[:\ddot{\underset{\cdot\cdot}{F}}\times]^{-}$	$3[Mg]^{2+}\,2[\times\ddot{\underset{\cdot\times}{P}}\times]^{3-}$

— Although an ionic compound is hard, because of the strong ionic bonds, it is also brittle, meaning it can be broken easily. Why? This is because when a force is applied across a plane of ions, it would displace it in such a way that the cations would now face one another while the anions would also face each other. This would repel the two planes a part, causing the lattice to crack.

(Continued)

(*Continued*)

— For a giant covalent compound, a similar tactic would not work as a much stronger force would be needed to break all the covalent bonds at one go, which is not an easy task.

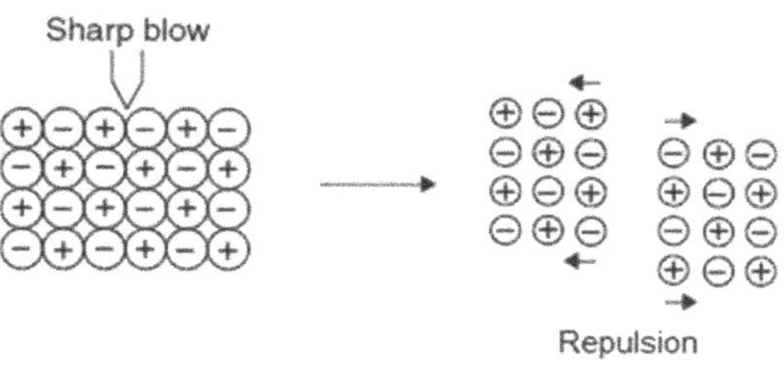

(iii) state the type of chemical bonding in the crystals; and

Explanation:

— For NaCl, it would be ionic bond and for SiO_2, it would be covalent bond.

Do you know?

— There is also instantaneous dipole–induced dipole interaction between the ions, but this interaction is too weak as compared to the strong ionic bond formed between the cation and anion. Therefore, we usually ignore it and focus more on the ionic bond.

(iv) explain why silicon dioxide adopts one type of bonding and sodium chloride another.

Explanation:

The electronegativity values for both Si and O do not differ too much. Hence, none of the atoms would prefer to lose electrons to the other. So, the most likely type of bonding would involve the sharing of electrons.

As for NaCl, the electronegativity values of Na and Cl differ too much, with Cl being very much more electronegative than Na. Hence, Na "does not mind" to lose an electron for Cl to gain. The formation of an cation and anion in NaCl, which have opposite charges, causes ionic bonding to ensue.

Do you know?

— Electronegativity is defined as the ability of the atom to distort shared electrons. This ability is correlated to the effective nuclear charge (ENC) that is acting on the valence electrons. The greater the ENC, the higher the electronegativity. Hence, electronegativity increases across a period but decreases down a group.

— The maxim 'metal reacts with non-metal to give ionic compound' works on this basis because metal has a much lower electronegativity value than a non-metal. But the interesting thing in chemistry is that there is bound to have some counter-examples such as $AlCl_3$, $PbCl_4$, and many more. But it is not difficult to understand why there are such counter-examples. It is simply because if the cation formed has very high charge density $\left(\propto \frac{q_+}{r_-}\right)$ and the anion formed has a very polarizable electron cloud, then the cation would pull the electron cloud from the anion to an extent that it causes the sharing of electrons to take place. The following diagram shows the various possible intermediate bond types:

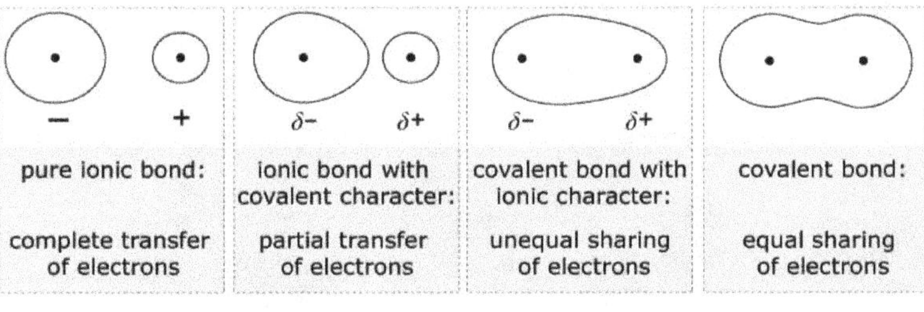

(*Continued*)

(*Continued*)

— Another way to understand why a particular type of compound is formed preferably, is to look at the energetics of its formation. A compound that is formed with the release of energy is more favorably formed than one that needs to take in energy. This is easy to understand as the release of energy lowers the energy level of the system, thus the system becomes more stable. Hence, when Si and O_2 are placed together, the formation of ionic SiO_2 would not be energetically favorable as compared to the formation of giant covalent SiO_2. As for ionic NaCl, they would not "like" to form covalent NaCl!

CHAPTER 3

IDEAL GAS AND GAS LAWS

1. (a) The critical point is a particular temperature and pressure condition whereby a supercritical fluid exists. A supercritical fluid is a special substance which is neither solid, liquid, nor gas, and carbon dioxide is one of the most widely used supercritical fluids. Calculate the volume of 1 mol of carbon dioxide at its critical point, assuming that it obeys the ideal gas equation.

   ($T_c = 304$ K, $p_c = 74 \times 10^5$ Nm^{-2}, $R = 8.31$ J K^{-1} mol^{-1})

Explanation:

Assuming that the supercritical fluid follows the ideal gas equation, $pV = nRT$:

$$\Rightarrow V = \frac{nRT}{p} = \frac{1 \times 8.31 \times 304}{74 \times 10^5} = 3.41 \times 10^{-4}\,m^3$$

Do you know?

— An ideal gas is one that follows the ideal gas equation religiously. Meaning for example, when p increases under constant n and T, V will decrease proportionately such that pV = nRT = constant.

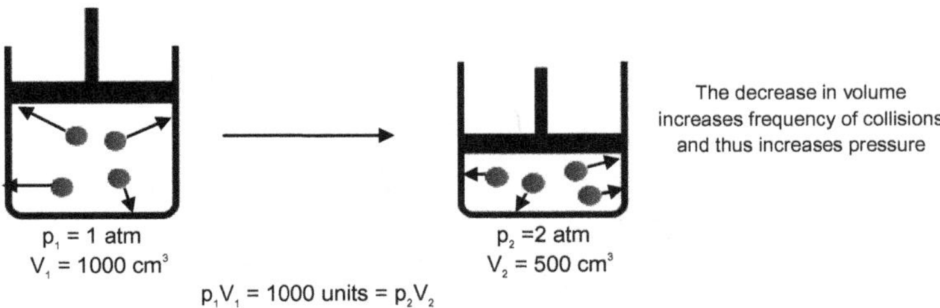

— Real gas does not follow the ideal gas equation because:

(i) Presence of intermolecular forces: When pressure increases on a real gas, the intermolecular forces would help to pull the gaseous particles closer together such that the actual volume decreases more than what the ideal gas equation would predict. That is, $pV < nRT$ than that for an ideal gas!

(ii) Particles have actual sizes: When pressure keeps on increasing, there would be a point in time where further increase in pressure would result in a decrease in volume that is much less than what is expected from the ideal gas equation. That is, $pV > nRT$ than that for an ideal gas!

In a nutshell, ideal gas does not exist in the real world! Because ideal gas particles do not have intermolecular forces and they exist as point mass, meaning they do not have a finite size. All these are present in real gas! The following graph shows the deviation of real gases from ideal

(*Continued*)

(*Continued*)

behavior under increasing pressure. And the stronger the intermolecular forces, the greater the amount of deviation from ideal behavior.

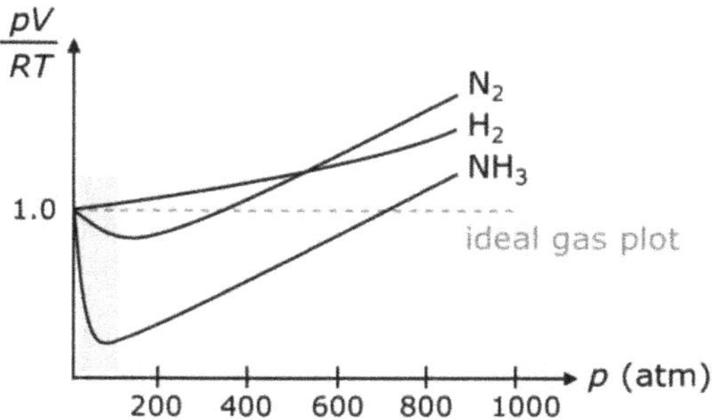

— But real gas can approximate ideal behavior, that is, it follows the pV = nRT equation, under:

(i) High temperature: At high temperatures, the gas particles have higher kinetic energy and hence move faster. As a result, the intermolecular forces become negligible as the particles do not "have time to attract each other."

(ii) Low pressure: When pressure is low, the particles are far apart, and as such, the intermolecular forces become negligible. In addition, the sizes of the particles are negligible as compared to the volume of the container.

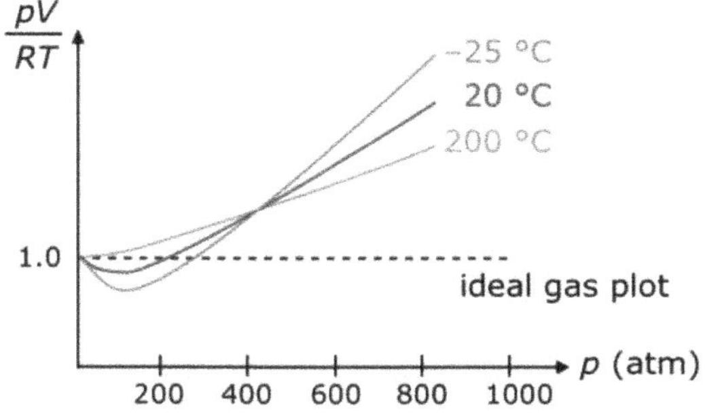

Q You said that a real gas does not follow the ideal gas equation because as pressure is increased on the gas, the volume does not decrease proportionately due to the presence of intermolecular forces, which pull the particles closer together. How would you use the same intermolecular forces idea to explain why pressure does not increase proportionately when the volume is decreased?

A: First, you need to understand what pressure is really about. Pressure is the collisional force exerted onto the wall of the container. The amount of collisional force acting on the wall depends on the number of gas particles in the system, the volume of the system, the mass of the gas particle, and the speed of the gas particle which is related to temperature.

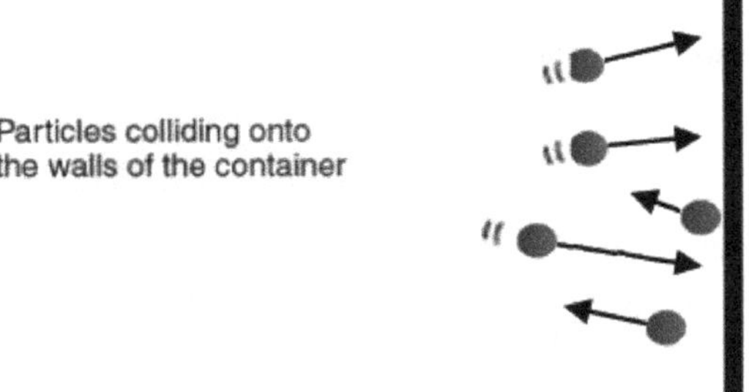

Particles colliding onto
the walls of the container

More gas particles at the same volume and temperature as compared to another would mean greater collisional frequency. Hence, higher pressure ($p \propto n$)!

A smaller volume at the same temperature would also mean greater collisional frequency as the particles have less space to move before knocking onto the wall of the container again. Hence, higher pressure ($p \propto 1/V$)!
A higher temperature with the same volume as compared to another would mean greater collisional frequency as the particles have more kinetic energy, hence they would move very fast. Therefore, the duration between two collisions is shorter. Hence, higher pressure ($p \propto T$)!

So, once you have understood what pressure is really about, imagine now you decrease the volume of the container, the immediate effect is that the particles have less space to move, thus they would collide more frequently with the wall of the container. But if there are intermolecular forces pulling on the particle that is about to knock onto the wall, wouldn't the intermolecular forces slow down the speed of the particle and hence the pressure, which is $\frac{\text{Force}}{\text{Area}}$? Thus, for a real gas, when the volume is decreased, the intermolecular forces would cause the measured pressure to be even lower than the expected value from the ideal gas equation, i.e., $pV < nRT$.

(b) The actual volume, at the critical point, of 1 mol of carbon dioxide is different from the value predicted by the ideal gas equation. Explain the given phenomenon.

Explanation:

The reason for the discrepancy is because there are still intermolecular forces of the instantaneous dipole–induced dipole type between the carbon dioxide molecules at the critical point. Hence, the gas does not follow the ideal gas equation, $pV = nRT$.

(c) Van der Waals' equation, $(p + a/V^2)(V - b) = RT$, can be used to account for the effect of gas imperfection.
(i) What physical meaning do these constants a and b have?

Explanation:

The presence of intermolecular forces causes the actual measured pressure that the gas particles exerted on the wall, to be less than predicted by the ideal gas equation. Hence, the constant a is used to compensate for the difference between the actual measured pressure and the one predicted by the ideal gas equation. We can thus perceive constant a as a measurement of the strength of the intermolecular forces.

Due to the finite dimension of the gas particles, the gas particles actually occupy some of the spaces of the container. Hence, the actual measured volume is larger than if the gas particles do not have any size and they do not occupy any space at all. Therefore, by subtracting the constant *b* from the volume V, it gives us a volume that ideal gas particles would actually occupy if they do not have any finite sizes. We can thus perceive constant *b* as a measurement of the size of the particles.

(ii) Explain how the constants would be different for gases such as hydrogen, nitrogen, and ammonia.

Explanation:

Since constant *a* is a measurement of the strength of the intermolecular forces, the stronger the intermolecular forces, the greater the value of *a*. Hence, the value of *a* would increase from hydrogen to nitrogen to ammonia. Both hydrogen and nitrogen are non-polar molecules possessing instantaneous dipole–induced dipole interaction. As the nitrogen molecule has more electrons than hydrogen, its electron cloud is more polarizable, hence having stronger id–id interaction. For ammonia, the largest *a* value is due to the stronger hydrogen bond present.

Since constant *b* is a measurement of the size of the particles, the bigger the particle, the greater the value of *b*. Hence, the value of *b* would increase from hydrogen to nitrogen to ammonia. The size of a molecule is dependent on the number of atoms that make up the molecule itself and the size of each individual atom. A nitrogen atom is bigger than a hydrogen atom, hence the molecular size would increase from hydrogen to nitrogen to ammonia.

Do you know?

- Fortunately there are intermolecular forces between gaseous particles as without them, there is no way to liquefy a gas. It is the intermolecular forces that pull the gas particles together under high-pressure conditions to form liquid.
- Fortunately gas particles have finite size. Or else we would not be able to see both liquid and solid states.

So it is actually incorrect to say that when liquid water is converted to water vapor, the heat energy is used to break the hydrogen bond?

A: Yes, it is incorrect to say that. We would prefer to say that the energy is used to *overcome* the bond.

(d) A sample of urine containing 0.120 g of urea, NH_2CONH_2, is treated with an excess of nitrous acid. The urea reacted according to the following equation:

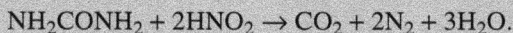

$$NH_2CONH_2 + 2HNO_2 \rightarrow CO_2 + 2N_2 + 3H_2O.$$

The gas produced is passed through aqueous potassium hydroxide and the final volume is measured. What is this volume at room temperature and pressure?

Explanation:

When the gas produced is passed through aqueous potassium hydroxide, the CO_2 reacts with the OH^- as follows:

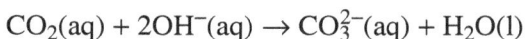

$$CO_2(aq) + 2OH^-(aq) \rightarrow CO_3^{2-}(aq) + H_2O(l)$$

$$\text{Molar mass of } NH_2CONH_2 = 14.0 + 2.0 + 12.0 + 16.0 + 14.0 + 2.0$$
$$= 60.0 \text{ g mol}^{-1}$$

Amount of $NH_2CONH_2 = 0.120/60.0 = 2.0 \times 10^{-3}$ mol

Amount of $N_2 = 2 \times 2.0 \times 10^{-3} = 4.0 \times 10^{-3}$ mol

Assuming ideal gas behavior, pV = nRT

$$\Rightarrow V = \frac{nRT}{p} = \frac{0.004 \times 8.314 \times 298}{101325} = 9.78 \times 10^{-5}\,m^3.$$

Do you know?

— When CO_2 is bubbled into water, the CO_2 is converted into carbonic acid, H_2CO_3.

$$CO_2(aq) + H_2O(aq) \rightarrow H_2CO_3(aq).$$

But if OH^- ion is present, the acidic H_2CO_3 will neutralize the OH^- to give the $CO_3^{2-}(aq)$ ion:

$$CO_2(aq) + 2OH^-(aq) \rightarrow CO_3^{2-}(aq) + H_2O(l).$$

If more CO_2 is bubbled into the solution, more H_2CO_3 will form. This would then react with the $CO_3^{2-}(aq)$ to give $HCO_3^-(aq)$:

$$CO_3^{2-}(aq) + H_2CO_3(aq) \rightarrow 2\ HCO_3^-(aq).$$

Thus, if you bubble too much CO_2 into $Ca(OH)_2(aq)$, you might miss the observation of the white precipitate (ppt), $CaCO_3$. This is because the more soluble $Ca(HCO_3)_2$ is formed instead!

— Another gas that behaves very similarly to CO_2 is actually SO_2. When SO_2 is bubbled into water, it forms weak sulfuric(IV) acid:

$$SO_2(aq) + H_2O(aq) \rightarrow H_2SO_3(aq).$$

If there is OH^- ion present, we would have:

$$SO_2(aq) + 2OH^-(aq) \rightarrow SO_3^{2-}(aq) + H_2O(l).$$

If more SO_2 is bubbled, we would have:

$$SO_3^{2-}(aq) + H_2SO_3(aq) \rightarrow 2\ HSO_3^-(aq).$$

Q When CO_2 is bubbled slowly into NaOH (aq), no white ppt is formed. Does that mean the CO_3^{2-}(aq) is not formed?

A: No! The CO_3^{2-}(aq) is still formed but because Na_2CO_3 is a soluble compound, you don't see the ppt being formed.

2. (a) (i) List the main assumptions of the kinetic theory for ideal gases.

Explanation:

The two main assumptions of the kinetic theory for ideal gases are:

(i) There are negligible intermolecular forces between the particles.
(ii) The volume of the particles is insignificant as compared to the volume of the container.

Do you know?

— There are actually five main assumptions made when scientists derived the pV = nRT equation mathematically from the basics. But only two are of interest to us in chemistry. The five assumptions are as follows:

(i) The particles exert no attractive or repulsive force on each other except during the process of collision. And between their collisions, the particles move in a straight-line trajectory.
(ii) The volume of the particles is small and they are very far apart. As a result, most of the volume of the container is actually empty space.
(iii) The particles are constantly in random motion. As a result, at any time, there is an equal number of particles moving in one direction as in another.
(iv) The particles are constantly colliding with each other and the wall of the container. The collision of the particles with the wall of the container gives rise to pressure.
(v) When particles collide with one another or the container, kinetic energy is conserved, i.e., the collision is elastic in nature. And because of this, the total kinetic energy of the system is constant unless energy is being transferred into or out of the system.

(ii) Suggest under what conditions real gases might deviate from ideality.

Explanation:

Real gases might deviate from ideality under:

(i) Low temperature: At low temperature, gas particles have lower kinetic energy and hence move slower. As a result, the intermolecular forces become prominent as the particles "have more time to attract each other."

(ii) High pressure: When pressure is high, the particles are very near to one another; as such, the intermolecular forces become prominent. In addition, the sizes of the particles are no longer insignificant as compared to the volume of the container.

(b) Experimentally, it has been found that the rate at which a gas diffuses through a porous barrier at constant temperature and pressure is inversely proportional to the square root of its molar mass.

$$\text{Rate} \ \alpha \ (1/M_r)^{1/2}$$

Carboxylic acids are known to dimerize in the gas phase. A sample of ethanoic acid vapor took twice as long to diffuse through a porous barrier as the same amount of neon gas. Use this information to calculate:

(i) the apparent molar mass of the ethanoic acid vapor, and

Explanation:

M_r of neon = 20.2

Rate of diffusion of neon = $R_{Ne} = \frac{k}{(20.2)^{\frac{1}{2}}}$ where k is the proportionality constant.

Since the time of diffusion of ethanoic acid is twice of that of neon, the rate of diffusion of ethanoic acid = $\frac{k}{(AMr)^{\frac{1}{2}}}$ = ½ of R_{Ne}

$$\Rightarrow \frac{k}{(AMr)^{\frac{1}{2}}} = \frac{1}{2}\frac{k}{(20.2)^{\frac{1}{2}}}$$

$$\Rightarrow AM_r = 4 \times 20.2 = 80.4$$

Therefore, apparent molar mass of ethanoic acid = 80.4 g mol^{-1}.

Q What is an apparent molar mass?

A: The word 'apparent' means that it is not the actual one. Molar mass is defined as the mass of ONE mole of particles. That means, this one mole of particles can actually consist of a composition of different types of particles, where each type of particles may have its own molar mass. Hence, if we take the total mass of this one mole of different types of particles, we would get the apparent molar mass. Mathematically, the apparent molar mass is defined as:

Apparent molar mass = $x_1M_1 + x_2M_2 + \ldots x_nM_n$ where x = mole fraction, and M = molar mass.

Q What is 'mole fraction'?

A: Mole fraction refers to the number of moles of a particular type of particle over the total number of moles of all the particles in the system. Since the quantity has the unit mole divided by another unit which is also mole, it is a dimensionless quantity!

(ii) the mole fraction of dimeric ethanoic acid in the vapor phase under these conditions.

Explanation:

Let the mole fraction of dimerized ethanoic acid be x.

Then, the mole fraction of ethanoic acid is $(1 - x)$.

M_r of ethanoic acid (CH_sCOOH) = 60.0 and dimerized ethanoic acid = 120.0

Hence, apparent molar mass $= 120x + 60(1 - x) = 80.4$

$$\Rightarrow x = 0.34$$

Therefore, the mole fraction of dimerized ethanoic acid is 0.34.

Do you know?

— The extent of dimerization is dependent on temperature and pressure. Hence, the higher the temperature, the smaller the mole fraction. This is because at a higher temperature, the bond that holds the dimer together is more likely to break. This is in accordance to the maxim, 'bond breaking needs energy.'

— At a higher pressure, the particles are being "forced" closer together. Hence, dimerization is more likely to take place.

(c) (i) Suggest how increasing the pressure and increasing the temperature will affect the rate of diffusion of neon.

Explanation:

If we assume that an increase in pressure here is due to a decrease in volume, then in a smaller volume, the neon particles are more likely to collide with the wall of the porous material. Hence, the increase in frequency of collision with the wall would make the neon particles more likely to pass through the porous material.

An increase in temperature causes the kinetic energy of the particles to increase. This thus increases the speed and hence increases the frequency of collision that the neon particles make with the wall. Thus, this causes the neon particles to be more likely to pass through the porous material.

Would an increase in temperature also cause the ethanoic acid vapor to diffuse faster?

A: Now, an increase in temperature decreases the amount of dimerization. Thus, we have more particles colliding with the wall of the porous material. Therefore, diffusion rate should increase. At the same time, the particles would also have more kinetic energy, which means a higher diffusion rate. In addition, a dimer which is heavier moves slower than a monomer. Thus, with more monomers, there would be a greater rate of diffusion. So you see, this is how we analyze the question by simply looking at how an increase in temperature is being translated into the factors that affect a molecule in the atomic world. And from there, we just simply treat the molecules as a random moving entity.

(ii) Assuming that only a small amount of gas is allowed to diffuse, explain how increasing the pressure and temperature would affect the rate of diffusion of the ethanoic acid vapor.

Explanation:

If we assume that an increase in pressure leads to more dimerization, then with a greater amount of heavier molecules, the diffusion rate will be slower. In addition, as dimerization increases, the number of particles also decreases; this would thus decrease the frequency of collision the particles make with the wall of the porous material. Hence, the diffusion rate decreases.

But if an increase in pressure does not lead to more dimerization, then the increase in the frequency of collision with the wall of the porous material would increase the diffusion rate.

An increase in temperature would certainly decrease the degree of dimerization. Hence, there are more particles that have lighter mass. This would increase the rate of diffusion due to the mass effect. At the same time, with more particles, there would be a greater frequency of collision with the wall of the porous material, hence a greater diffusion rate. In addition, with a higher temperature, the particles would also have more kinetic energy. There would be a greater frequency of collision with the wall of the porous material, which means a higher diffusion rate.

Q What is partial pressure of a gas?

A: We have defined pressure simply as the collisional force per unit area that the particles collide on the wall of the container. Thus, if we have a mixture of gas particles, each of the different types of gas particles would collide INDEPENDENTLY on the wall of the container. So, according to the ideal gas equation, the total pressure of the system is:

$$p_TV = (n_1+n_2+\ldots n_x)RT = n_1RT + n_2RT + \cdots + n_xRT$$

$\Rightarrow p_T = \frac{RTn_1}{V} + \frac{RTn_2}{V} + \ldots + \frac{RTn_x}{V} = p_1 + p_2 + \ldots + p_x$, where $p_1, p_2, \ldots$ are partial pressures.

The above equation is actually Dalton's Law of Partial Pressure.

Now, since $p_1 = \frac{RTn_1}{V} = \frac{RTn_1}{V} \times \frac{n_T}{n_T} = x_1 \frac{RTn_T}{V}$, where $x_1 = \frac{n_1}{n_T}$ is the mole fraction,

$$\Rightarrow p_1 = x_1 \cdot p_T$$

That is, the partial pressure of a gas is actually a fraction of the total pressure. And under constant V and T, the partial pressure of a gas is directly proportional to the number of moles of the particles, i.e., $p \propto n$.

3. Gaseous xenon tetrafluoride, XeF_4, at a partial pressure of 2.0 kPa, and hydrogen, at a partial pressure of 10 kPa, are exploded in an enclosed container producing xenon and hydrogen fluoride.
   (a) Write the balanced equation for the reaction between xenon tetrafluoride and hydrogen.

Explanation:

$$XeF_4(g) + 2H_2(g) \rightarrow Xe(g) + 4HF(g)$$

(b) Calculate, in terms of partial pressure, how much of hydrogen reacted with all the xenon tetrafluoride present.

Explanation:

Under constant T and V, $p \propto n$. Therefore, we can consider the reacting mole ratio in terms of ratio of the partial pressures of the reactants.

Hence, 1 kPa of $XeF_4(g)$ reacts with 2 kPa of $H_2(g)$.

Since the initial partial pressure of $XeF_4(g)$ is 2.0 kPa, 4.0 kPa of $H_2(g)$ is needed.

(c) Hence, calculate the total pressure of the mixture of gases in the enclosed container, assuming that the temperature remains constant.

Explanation:

Partial pressure of: remaining $H_2 = 10 - 4 = 6.0$ kPa
$Xe = 2.0$ kPa
$HF = 4 \times 2.0 = 8.0$ kPa

Total pressure = sum of all the partial pressures = 16.0 kPa.

(d) Suggest reasons why fluorides of xenon have been synthesized whereas fluorides of the other noble gases, e.g., helium, neon, and argon, are relatively uncommon.

Explanation:

Xenon has a greater atomic size than the rest of the noble gases. As such, the valence electrons are less strongly attracted and hence need a smaller amount of energy to be promoted to higher-energy vacant orbitals so as to form bonds with the fluorine atoms. This energy needed is later compensated by the energy being released when the Xe–F bonds formed.

Do you know?

— The whole issue of about asking whether a chemical reaction would proceed is equivalent to the investment-return theory in economics. If the investment is small but it reaps good returns, the investment is likely to take place. So, what is the investment in a chemical reaction? Well, it is the energy needed to break bonds or in this case here, to promote electrons from a lower energy level to higher ones. But what are the returns? It is the energy that would be released when bonds form. And the stronger the bonds that form, the greater the energy that would be released. Thus, in this case, the fact that XeF_4 is likely to form from Xe and F_2 must means that the energy that is released when the Xe–F bonds formed is able to compensate for the amount of energy needed for Xe and F_2 to react.

So the reason ArF_4 does not form is because the so-called 'investment return' sum is not right?

A: Yes, you can perceive it as that, although we would expect that the Ar–F bond that is formed to be stronger than the Xe–F bond, meaning we may get more energy "return." But as for the energy that is needed to bring valence electrons (which are more strongly attracted in Ar than Xe) into a higher energy level, this "investment" may not be able to be compensated by the "return."

4. A plot of ρ/p against pressure, p, at a given temperature for an unknown gas X is shown below.

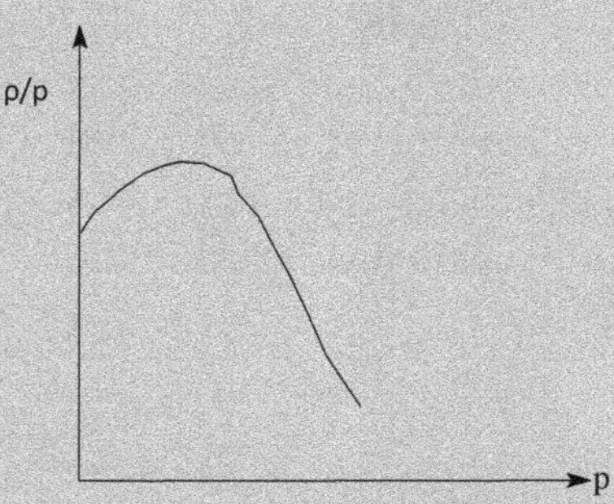

(a) Is the gas behaving ideally?

Explanation:

First, we need to understand what is $\frac{\rho}{P}$, where ρ is density of the gas.

$$\text{From } pV = nRT = \left(\frac{\text{mass}}{\text{molar mass}}\right)RT,$$

$$\Rightarrow p\ (\text{molar mass}) = \left(\frac{\text{mass}}{V}\right)RT = \rho RT$$

$$\Rightarrow \frac{\rho}{P} = \left(\frac{\text{molar mass}}{RT}\right) = \text{constant for an ideal gas.}$$

Since $\frac{\rho}{P}$ is not a constant when increasing p, the gas does not behave ideally.

(b) Give an explanation for the graph obtained.

Explanation:

Since, $\frac{\rho}{\rho} = \frac{\text{mass}}{pV}$, as pressure increases, the presence of the intermolecular forces would "help" to pull the gas particles even closer together. This makes the measured V decrease much more than what the ideal gas equation would predict. As such, the actual pV product is less than that for an ideal gas. Hence, $\frac{\rho}{p}$ increases initially.

As pressure keeps on increasing, there would be a point in time when a further increase of pressure would cause the measured V to decrease much less than what the ideal gas equation would predict. This is because the volume of the gas particles is no longer insignificant. As such, the actual pV product is more than that for an ideal gas. Hence, $\frac{\rho}{p}$ decreases at the higher pressure values.

(c) Using the axes provided above, give a sketch of ρ/p against pressure, p, for gas X under a higher temperature.

Explanation:

Under a higher temperature, the gas particles have more kinetic energy, hence the intermolecular forces become negligible as the particles do not "have time to attract one another." The gas would approach ideal behavior and hence there would be less deviation.

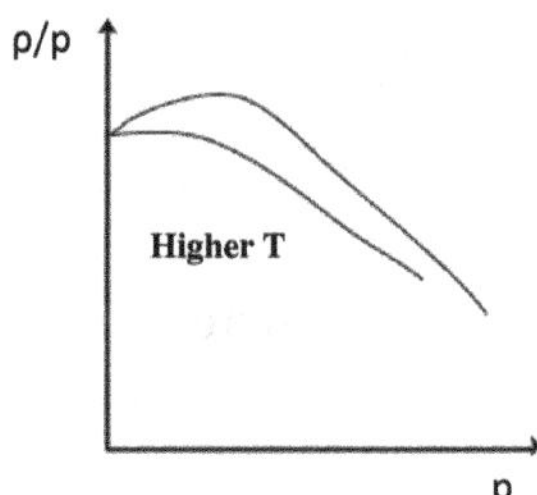

(d) Using the axes provided above, give a sketch of ρ/p against pressure, p, for gas Y, which has a higher boiling point than X.

Explanation:

Since gas Y has a higher boiling point than gas X, this would mean that the intermolecular forces for gas Y are stronger than those for gas X. Hence, gas Y would deviate more from ideal behavior than X.

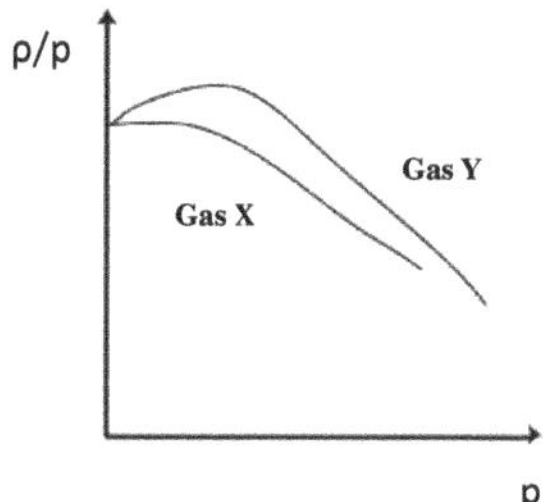

5. A sample of pure organic liquid **W** of mass 0.146 g is vaporized in a gas syringe at 127°C and occupied a volume of 100 cm^3 at a pressure of 101 kPa.
   (a) Calculate the relative molecular mass of **W** from the experimental data.

Explanation:

Mass of the gas = Mass of liquid
Assuming ideal behavior, $pV = nRT = \left(\frac{\text{mass}}{\text{molar mass}}\right)RT$

$$\Rightarrow \text{Molar mass} = \frac{8.314 \times 400 \times 1.46 \times 10^{-4}}{1 \times 10^{-4} \times 1.01 \times 10^{5}} = 0.0481\ \text{kg mol}^{-1} = 48.1\ \text{kg mol}^{-1}$$

Hence, the relative molecular mass of **W** is 48.1.

Do you know?

— Other than using mass spectrometry to determine the molecular mass of an unknown compound that can be vaporized, you can also use the pV = nRT equation to find it. Isn't it cool and simple?

(b) The accurate composition by mass of **W** is C, 52.2%; H, 13%; and O, 34.8%. Determine the empirical formula of **W**.

Explanation:

To calculate the empirical formula:

	C	H	O
Percentage by mass	52.2	13	34.8
No. of mole in 100 g	4.35	13	2.18
Mole ratio	2	6	1

The empirical formula is C_2H_6O.

(c) What are the likely formula and relative molecular mass of **W** based on part (*b*)?

Explanation:

Since n × empirical formula = molecular formula,

$$\Rightarrow n \times (24 + 6 + 16) = n \times 46 = 48.1.$$

The best value for n is 1, which means the molecular formula is also C_2H_6O.

So, the theoretical relative molecular mass for **W** is 46.0.

(d) Suggest two possible reasons for the difference between the values of the relative molecular masses in parts (*a*) and (*c*).

Explanation:

The relative molecular mass in part (*a*) was obtained using the ideal gas equation, which assumed that gas **W** does not have (i) intermolecular forces and (ii) finite sizes. These two assumptions are obviously invalid as gas **W** is not an ideal gas at all. Thus, the measured volume for gas **W** is inaccurate because of these two reasons. In fact, the measured volume of 100 cm^3 is smaller than what the ideal gas equation would predict.

Do you know?

— Some of the important graphs that you need to know:
At constant T,

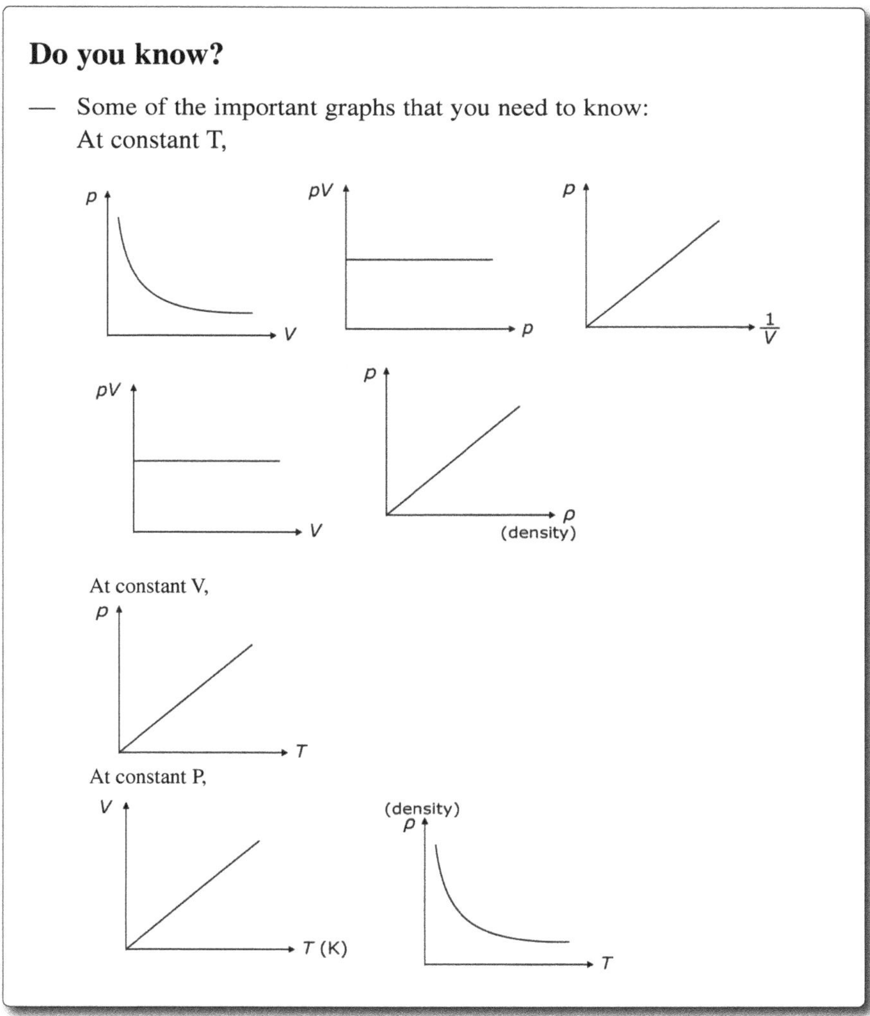

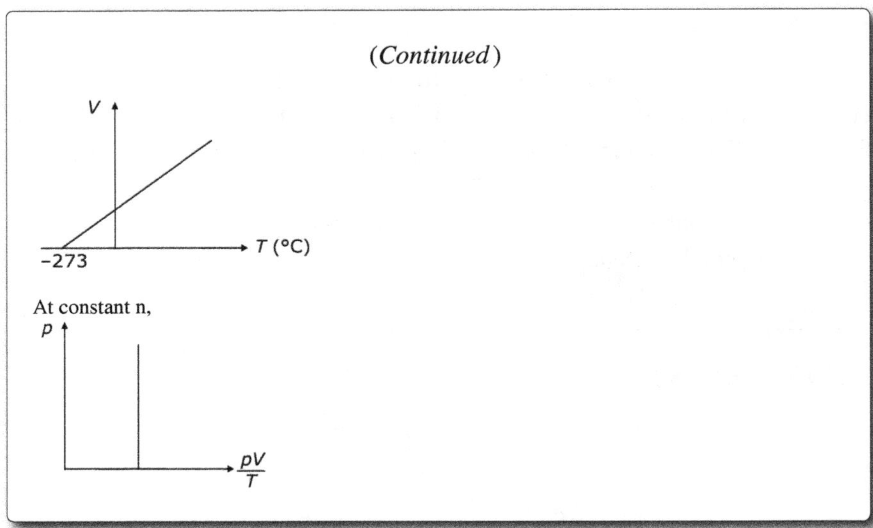
(Continued)
V
T (°C)
−273
At constant n,
p
pV/T

CHAPTER 4

CHEMICAL THERMODYNAMICS

1. (a) During physical training, people may suffer from twisted ankles. In such cases, ice should be applied to shrink the blood vessels around the sprain in order to minimize any internal bleeding. But storing ice for such usage is neither convenient nor economical. So, trainers often use cold packs consisting of a divided plastic bag containing ammonium nitrate and water. (Given that the $\Delta H^{\theta}{}_{sol}(NH_4NO_3) = + 26.0$ kJ mol^{-1}.)

   (i) Determine the heat change when 30.0 g of ammonium nitrate dissolves in water.

Explanation:

$NH_4NO_3(s) + aq \rightarrow NH_4NO_3(aq)$ $\quad \Delta H^{\theta}{}_{sol}(NH_4NO_3) = +26.0$ kJ mol^{-1}.

Molar mass of $NH_4NO_3(s) = 80.0$ g mol^{-1}

Amount of NH_4NO_3 in 30.0 g = 30.0/80.0 = 0.375 mol

Heat change = 26.0 x 0.375 = + 9.75 kJ.

Do you know?

— The value for heat change must carry a sign of '+' or '–' to indicate whether heat energy is absorbed (endothermic) or released (exothermic). An exothermic reaction would indicate that the products have a lower energy state than the reactants, hence the products are energetically more

(Continued)

(Continued)

stable than the reactants, and vice versa! The energy profile diagrams below show it:

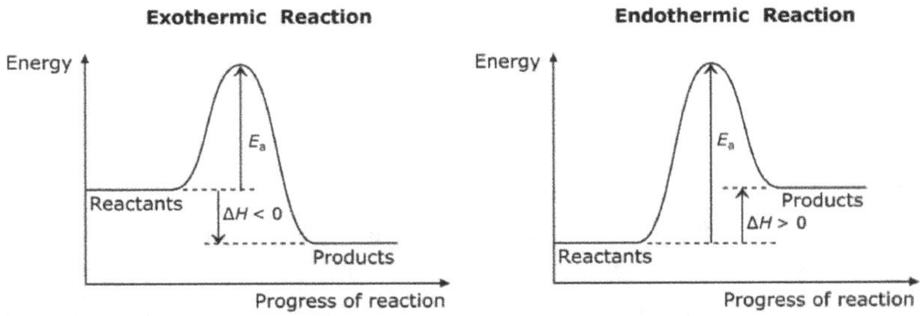

In addition, an exothermic enthalpy change would also indicate that the bonds in the products are stronger than bonds in the reactants!

— The enthalpy change of solution for an ionic compound is as follows:

$$\Delta H^{\theta}_{sol} = \Delta H^{\theta}_{hyd} - \text{L.E.},$$

where $\Delta H^{\theta}_{hyd} = \Delta H^{\theta}_{hyd}(\text{cation}) + \Delta H^{\theta}_{hyd}(\text{anion})$

$$\text{L.E.} = \text{Lattice energy} \propto \frac{q_+ \cdot q_-}{(r_+ + r_-)}$$

Thus, if an ionic compound has weak ionic bonding (indicated by less exothermic L.E.) and a very exothermic enthalpy change of hydration, the ionic compound is more likely to be soluble as the ΔH^{θ}_{sol} is more likely to be exothermic.

Although the ΔH^{θ}_{sol} of NH_4NO_3 is endothermic, it is still soluble. So, does it mean that as long as ΔH^{θ}_{sol} is endothermic in value, it would indicate that the compound is soluble?

A: NO! In fact, to predict whether a compound is soluble or generally whether a reaction is feasible, we need to use $\Delta G^{\theta} = \Delta H^{\theta} - T\Delta S^{\theta}$ to predict. The following table summarizes all the possibilities:

ΔH^{θ}	ΔS^{θ}	ΔG^{θ}	Thermodynamically feasible?
+	+	– if $\lvert T\Delta S^{\theta}\rvert > \lvert \Delta H^{\theta}\rvert$	At high T
+	–	Always +	Not at all
–	+	Always –	At all T
–	–	– if $\lvert T\Delta S^{\theta}\rvert < \lvert \Delta H^{\theta}\rvert$	At low T

So a negative ΔG^{θ} would mean that the reaction is thermodynamically feasible as predicted. But this does not mean that the reaction would certainly happen as there are other factors, such as high activation energy, that prevent the reaction from occurring under standard conditions. Take for instance, diamond is actually less stable than graphite from the enthalpy perspective. But do you have diamond turning into graphite instantaneously? No! Right? This is because the conversion from diamond to graphite has a high activation energy!

When an ionic compound dissolves, free ions are formed, hence the system is more disordered. So is the ΔS^{θ}_{sol} always positive?

A: Not necessarily! Dissolved ions are more disordered than the solid ionic compound. But do not forget that the ions are surrounded by layers of water molecules bonded via the ion–dipole interaction. These strongly bonded water molecules have decreased the degree of disorderliness of the system, hence the entropy. Thus, if the entropy of the water molecules decreases much more than the increase in entropy of the ions, then overall, the ΔS^{θ}_{sol} of the system would be negative. But if it is the other way round, then ΔS^{θ}_{sol} would be positive.

Q So, how would I know whether the ΔS^{θ}_{sol} is positive?

A: Elementary! If we tell you that the solution process is spontaneous, which means the ΔG^{θ}_{sol} is negative and the ΔH^{θ}_{sol} is positive, what is your best guess from the $\Delta G^{\theta} = \Delta H^{\theta} - T\Delta S^{\theta}$ equation?

Q For a negative ΔG^{θ}, do we say that the reaction is exothermic?

A: No! Exothermic or endothermic is only meant for ΔH^{θ}. If ΔG^{θ} is negative, we say that the reaction is *exergonic* while if it is positive, it is *endergonic*.

Do you know?

— The factors that affect entropy includes:

(1) Temperature: An increase in temperature increases the entropy as the particles in the system have more energy to distribute into the various energy states, therefore degree of disorderliness increases;
(2) Phase change: Both melting and vaporization result in an increase of the entropy as the particles in the liquid state are more disordered than in the solid state; and the gas particles are more disordered than those in the liquid state (i.e., $S_{gas} \gg S_{liquid} > S_{solid}$);
(3) Number of particles: As the number of particles in a system increases, the degree of disorderliness increases; and
(4) Mixing: If mixing takes place, the degree of disorderliness in the system increases because each particle has more space to move about.

(ii) What would be the final temperature if 30.0 g of ammonium nitrate is added to 200 g of water at 298 K? Assume that the heat capacity of ammonium nitrate solution is 4.0 J g^{-1} K^{-1}.

Explanation:

$$\text{Heat change} = mc\Delta T = 200 \times 4.0 \times \Delta T = 9.75 \times 10^3$$
$$\Rightarrow \Delta T = -12.2^{\circ}C$$
$$\Delta T = T_f - T_i = T_f - 298 = -12.2$$
$$\Rightarrow T_f = 286.2 \text{ K}$$

Do you know?

— Change is always calculated by taking the final minus the initial. Hence, a change must carry a '+' or '−.'

(*Continued*)

(Continued)

— Sometimes, you may be given the volume of the solution rather than the mass of the water. But when you calculate the heat change using $mc\Delta T$, you still need to substitute the m by using the volume of the solution. Why? This is because when heat is released, it goes as kinetic energy of the particles in the system. Now, since water is present in abundance, they are the particles that "shoulder" most of the heat energy. Vice versa when the reaction absorbs heat energy, most of the energy is supplied from the kinetic energy of the water molecules. In addition, if we take the density of a solution to be equivalent to that of pure water, i.e., 1.0 g cm^{-3}, then:

$$m = \rho \times \text{volume} = \text{volume}.$$

(b) Ammonium nitrate is widely used as an explosive because it can decompose according to the following equation:

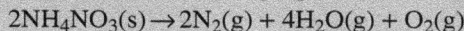

$$2NH_4NO_3(s) \rightarrow 2N_2(g) + 4H_2O(g) + O_2(g)$$

(i) With reference to the Data Booklet, determine the enthalpy change for this decomposition. (Assume that the L.E. of NH_4NO_3 is −653 kJ mol^{-1}.)

Explanation:

The energy cycle is as follows:

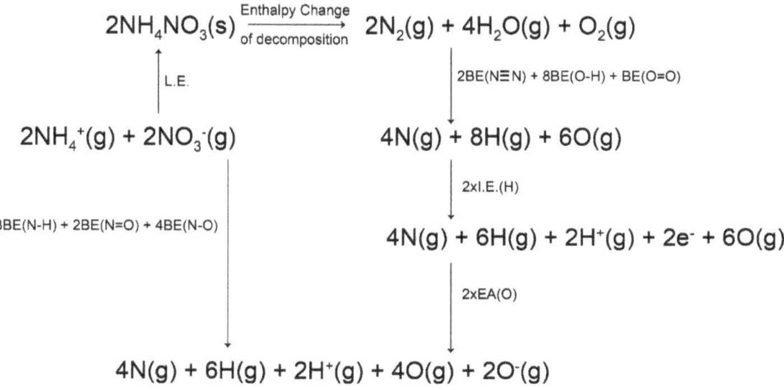

By Hess' Law,

$$\begin{aligned}\Delta H_{decomp} &= -\text{L.E.} + \{8\text{BE(N–H)} + 2\text{BE(N=O)} + 4\text{BE(N–O)}\} - \\ &\quad 2\text{EA(O)} - 2\text{I.E.(H)} - \{2\text{BE(N}\equiv\text{N)} + 8\text{BE(O–H)} + \\ &\quad \text{BE(O=O)}\} \\ &= -(-653) + 8(390) + 2(590) + 4(222) - \\ &\quad 2(-(141)) - 2(1310) - \{2(994) + 8(460) + 496\} \\ &= -3225 \text{ kJ mol}^{-1}.\end{aligned}$$

Q What is Hess' Law?

A: Hess' Law is actually an expression of the more general Law of Conservation of Energy, which states that energy cannot be created nor destroyed but transferred. So, Hess' Law states that the enthalpy change for a chemical process is only dependent on the initial and final states but independent the pathway taken.

Do you know?

— When you construct an energy cycle, just make sure you go according to the definition of the enthalpy terms. Do not unnecessarily change the direction of the energy change. The following procedure would be useful:

Step 1:

An energy cycle looks like this:

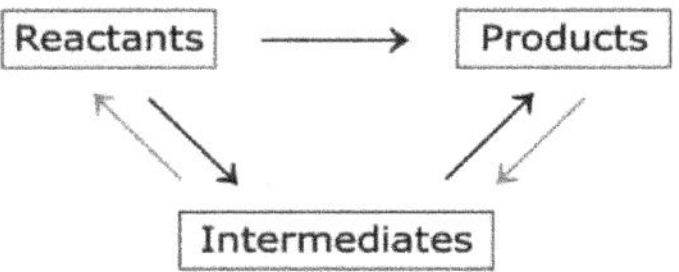

Convert the enthalpy change in question into an equation, i.e., *Reactants* → *Products*. This is the first pathway created. (Note: It is helpful to know the definitions well.)

(*Continued*)

(*Continued*)

Step 2:

Based on the enthalpy changes that are given in the question, create an alternative pathway by filling up *Intermediates*.

For example, if ΔH_f data are given, then *Intermediates* must be the elements, with arrows pointing from *Intermediates* to both *Reactants* and *Products*; if ΔH_c data are given, then the arrows must be pointing from both *Reactants* and *Products* to *Intermediates*.

Step 3:

Once the cycle is completed, apply Hess' Law to sum up the enthalpy values for the two pathways. But we normally use the point where the enthalpy value is to be determined as the initial state and in this case here, it is the NH_4NO_3 state. So, there would always be two different pathways, one clockwise while the other is anti-clockwise. All you need to do is when you come across an enthalpy term that is anti to the direction that you are moving in, reverse the sign of this enthalpy term by taking the negative of it! Note that any point in the cycle can be the starting point. In addition, remember to account for material balance!

(ii) What features of its decomposition make ammonium nitrate explosive?

Explanation:

The decomposition process is highly exothermic, releasing much heat energy in the form of the kinetic energy of the gas particles. In addition, since many gas particles are produced, these highly energetic, gas particles result in an explosive nature for the decomposition.

Do you know?

— If you calculate the ΔG^{θ} of such an explosive reaction using the $\Delta G^{\theta} = \Delta H^{\theta} - T\Delta S^{\theta}$ equation, you would notice that the ΔG^{θ} is a highly negative value, which indicates that the reaction is very thermodynamically spontaneous. And the contributory factors are actually due to the highly exothermic ΔH^{θ} and the very positive ΔS^{θ} as all the products are gas particles which have a very high degree of disorderliness.

2. (a) With reference to the Data Booklet, calculate the standard enthalpy change of the formation of calcium(I) chloride. (Assume $CaCl(s) \rightarrow Ca^{+}(g) + Cl^{-}(g)$).

Explanation:

The Born–Haber cycle for the formation of CaCl(s) is shown below:

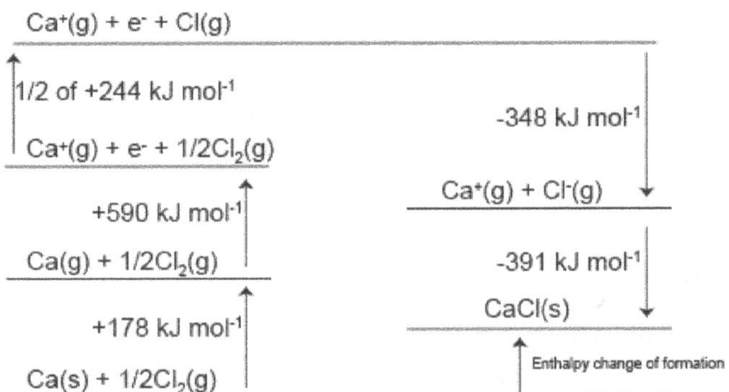

By Hess' Law,

$$\begin{aligned}\Delta H_f(CaCl(s)) &= \Delta H_{atm}(Ca(s)) + 1^{st}\ I.E.(Ca) + \Delta H_{atm}(Cl) \\ &\quad + EA(Cl) + L.E.(CaCl(s)) \\ &= 178 + 590 + \tfrac{1}{2}(244) - 348 - 391 = +151\ kJ\ mol^{-1}.\end{aligned}$$

Do you know?

— The Born–Haber cycle is a special type of energy level diagram used to depict the formation of an ionic compound. The important features of an energy level diagram is that (i) an endothermic reaction must be shown by a vertical arrow going upward, and (ii) an exothermic reaction must be shown by a vertical arrow going downward. A typical energy level diagram is shown below:

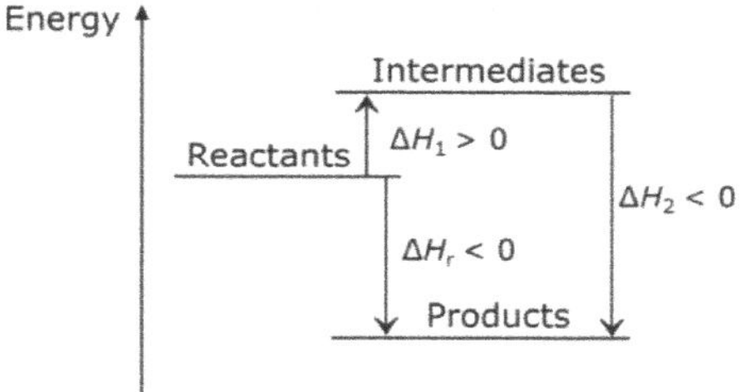

— Enthalpy change of atomization of Ca(s) is also equivalent to the enthalpy change of sublimation of Ca(s). It is a measurement of the metallic bond strength of Ca(s).

— Enthalpy change of atomization of Cl_2 is numerically equivalent to half the bond energy of Cl_2. This is because according to the definition of enthalpy change of atomization, which is to form one mole of gaseous atoms, the energy needed would be half of the amount of energy required to break one mole of Cl_2 molecules (definition of bond energy), forming two moles of Cl atoms.

Since ΔH_f(CaCl(s)) is a positive value, does it mean that CaCl(s) is less likely to form?

A: CaCl(s) does not exist but $CaCl_2$(s) does! The enthalpy change of formation of $CaCl_2$(s) is -795 kJ mol^{-1}, an indication that the formation is energetically favorable.

But in order to form $CaCl_2(s)$, we need more energy in terms of using it to remove the 2nd electron from Ca(g) and to form more Cl(g) atoms. So, shouldn't these make the formation of $CaCl_2(s)$ less likely?

A: You are right that we need to use more energy to form the $Ca^{2+}(g)$ ion and to form an additional mole of Cl(g), but we reap a better "return" when the solid $CaCl_2(s)$ forms. This is in terms of the more exothermic L.E., $Ca^{2+}(g) + 2Cl^-(g) \rightarrow CaCl_2(s)$, that is released. It is about -2195 kJ mol^{-1}. This is in line with the investment return theory in economics that we have discussed previously and Nature is pretty good at it!

Why is the L.E. of $CaCl_2(s)$ so much exothermic than that of CaCl(s)?

A: Well, remember that L.E. $\propto \frac{q_+ \cdot q_-}{(r_+ + r_-)}$? $CaCl_2$ consists of Ca^{2+} which has a higher charge and smaller cationic radius than Ca^+, thus the L.E., which is also a measurement of the ionic bond strength, is more exothermic.

(b) The standard enthalpy change of formation of calcium(II) chloride is -795 kJ mol^{-1}. Calculate the enthalpy change for the reduction of calcium(II) chloride to calcium(I) chloride by calcium metal.

Explanation:

$\Delta H_f(CaCl_2(s)) = -795$ kJ mol^{-1} $\quad Ca(s) + Cl_2(g) \rightarrow CaCl_2(s)$ (1)

$\Delta H_f(CaCl(s)) = +151$ kJ mol^{-1} $\quad Ca(s) + 1/2Cl_2(g) \rightarrow CaCl(s)$ (2)

ΔH_{Red}? $\quad CaCl_2(s) + Ca(s) \rightarrow 2CaCl(s)$ (3)

Equation (3) = 2 × (2) – (1) i.e., $\Delta H_{Red} = 2(151) - (-795) = +1097$ kJ mol^{-1}.

Wow! I didn't know that we can solve enthalpy change simultaneously just like solving linear equations!

A: Well, as long as the materials are balanced for all the equations, you can do it.

(c) Experimentally, making calcium(I) chloride by simply reducing calcium(II) chloride with calcium metal has not been fruitful. Theoretically, is it possible to make calcium(I) chloride by this method? Explain.

Explanation:

The $\Delta H_{\mathrm{Red}} > 0$ while the $\Delta S_{\mathrm{Red}} \approx 0$, hence $\Delta G_{\mathrm{Red}} = \Delta H_{\mathrm{Red}} - \mathrm{T}\Delta S_{\mathrm{Red}} \approx \Delta H_{\mathrm{Red}} > 0$. Since $\Delta G_{\mathrm{Red}} > 0$, theoretically, the formation of calcium(I) chloride from this method is thermodynamically non-spontaneous.

Do you know?

— An Ellingham diagram graphically depicts the thermodynamic spontaneity of a reaction with respect to ΔG. The following Ellingham diagram is for the reduction of metal oxides:

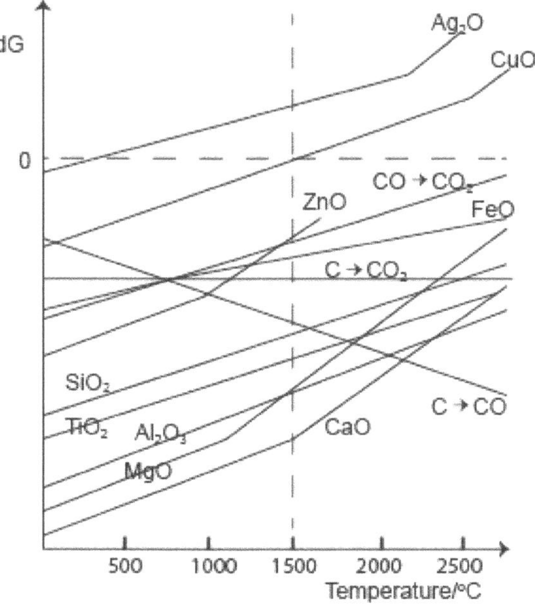

(*Continued*)

(Continued)

- As temperature increases, ΔG of formation of a particular oxide (except for CO and CO_2) increases according to $\Delta G = \Delta H - T\Delta S$, if we assume that both ΔH and ΔS are invariant with temperature. But for some compounds, their gradient change after a particular temperature. For example, there is a kink at around 1500°C for CaO. This phenomenon arises because at 1500°C, the CaO changes phase. That means that the type of CaO is different before and after 1500°C in terms of crystalline structure.
- The gradient of the ΔG of formation versus temperature plot gives us the ΔS of the formation in accordance to $\Delta G = \Delta H - T\Delta S$.
- As temperature increases, the ΔG of formation becomes less negative or more endergonic. This is an indication that the stability of the metal oxide decreases with respect to the increasing temperature.
- Carbon, which is the main component of coal, is a common reducing agent that can reduce a lot of metal oxides, such as FeO, ZnO, CuO, and Ag_2O, back to their metallic form. The reducing power of carbon increases with temperature as indicated by the more negative ΔG.

Q How would we know whether a reduction of the metal oxide would take place? For example, between Al(s) and FeO(s)?

A: Let us assume the temperature is 1500°C:

$2Al(s) + \frac{3}{2}O_2(g) \rightarrow Al_2O_3(g) \quad \Delta G_f(Al_2O_3) = -b \text{ kJ mol}^{-1}$
$Fe(s) + \frac{1}{2}O_2(g) \rightarrow FeO(s) \quad \Delta G_f(FeO) = -a \text{ kJ mol}^{-1}$
$2Al(s) + 3FeO(s) \rightarrow 3Fe(s) + Al_2O_3(s)$
The $\Delta G_{Red} = \Delta G_f(Al_2O_3) - 3\Delta G_f(FeO) = (-b + 3a) \text{ kJ mol}^{-1}$

Thus, if the calculated $\Delta G_{Red} < 0$, then the reduction process would be thermodynamically spontaneous. Basically, the idea is very simple. If you want to use metal 1 to convert the oxide of another metal 2 into metal 2, then all you need to do is look for a metal 1 that has a ΔG value that is lower than the ΔG of the oxide of metal 2.

What happens when two of the lines intersect each other in the Ellingham diagram?

A: If the two lines intersect each other, this would mean that the overall ΔG for the reduction process is zero. The system would reach an equilibrium state! The reduction reaction would not go to completion. If we look at the intersection point for both MgO and Al_2O_3 at 1500°C, below this temperature, Mg(s) would be able to reduce Al_2O_3(s). But above 1500°C, it is Al(s) that is able to reduce MgO(s).

3. With a dwindling reserve of fossil fuels, scientists have suggested using methanol (CH_3OH) as a possible alternative fuel for motor cars. The advantages of using methanol as a fuel are that it burns cleanly, giving out less pollutants than gasoline, and is less of a fire hazard in an accident.
   (a) Using your Data Booklet, calculate the enthalpy change when one liter of methanol is burned in excess oxygen. (Assume the density of methanol is 0.79 g cm^{-3} and ΔH_{vap}(methanol) = +38.3 kJ mol^{-1}.)

Explanation:

Molar mass of CH_3OH = 32 g mol^{-1}.
Amount of CH_3OH = (0.79 × 1000)/32 = 24.69 mol.

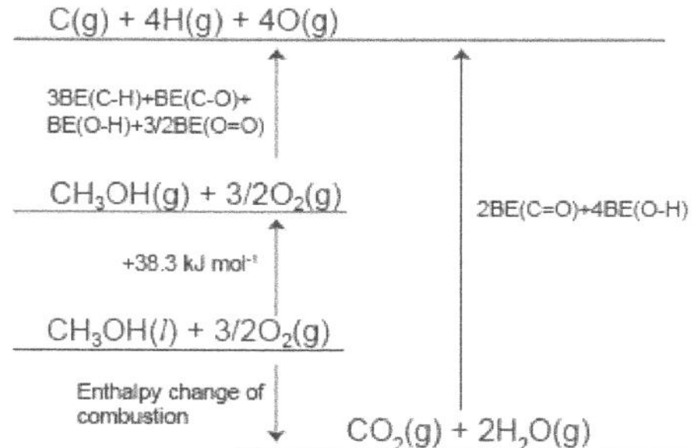

By Hess' Law,

$$\begin{aligned}\Delta H_{comb} &= \Delta H_{vap}(\text{methanol}) + 3BE(C\text{–}H) + BE(C\text{–}O) + BE(O\text{–}H) \\ &\quad + 3/2BE(O{=}O) - [2BE(C{=}O) + 4BE(O\text{–}H)] \\ &= 38.3 + 3(410) + 360 + 460 + 3/2(496) - [2(740) \\ &\quad +4(460)] \\ &= -487.7 \text{ kJ mol}^{-1}\end{aligned}$$

Thus, the enthalpy change for 1 liter of methanol being combusted = 24.69 × (−487.7) = −12041.3 kJ.

(b) The heat change on running one liter of petrol is approximately −33000 kJ. Comment on the significance of your answer in part *(a)* to the design and operation of cars which run on methanol.

Explanation:

Since one liter of methanol does not evolve as much heat energy as a liter or petrol, for the car to travel the same distance, it needs to carry more methanol fuel. Hence, the fuel tank probably has to be a bigger one. If not, the car needs to be smaller so that it does not need too much energy to move. Otherwise, a methanol-operating car is probably only suitable for short-distance driving.

(c) Explain why methanol is less fire hazardous.

Explanation:

Methanol is less volatile than petrol, hence less likely to be ignited. In addition, the combustion of methanol does not release as much energy as petrol, so it is relatively safer.

4. (a) Define the term standard enthalpy change of formation.

Explanation:

The energy change when 1 mole of a pure compound in a specified state is formed from its constituent elements in their standard states, at 298 K and 1 bar.

Do you know?

— *Standard enthalpy change of reaction*: The energy change when molar quantities of reactants, as specified by the chemical equation, react to form products at 298 K and 1 bar.

— *Standard enthalpy change of combustion*: The energy *evolved* (i.e., exothermic) when 1 mole of a substance is completely burned in excess oxygen at 298 K and 1 bar.

— *Bond energy*: The bond energy of an X–Y bond is the average energy *absorbed* when 1 mole of X–Y bonds are broken in the gaseous state.

— *Bond dissociation energy*: Bond dissociation energy of an X–Y bond is the energy *absorbed* to break 1 mole of that particular X–Y bond in a particular compound in the gaseous state.

— *Lattice energy*: The energy *evolved* when 1 mole of a pure solid ionic compound is formed from its constituent gaseous ions.

— *1st ionization energy*: The energy *absorbed* to remove 1 mole of electrons from 1 mole of gaseous atoms in the ground state to form 1 mole of gaseous X^+ ions.

— *First electron affinity*: The energy change when 1 mole of electrons is added to 1 mole of gaseous atoms to form 1 mole of gaseous X^- ions.

— *Standard enthalpy change of solution*: The energy change when 1 mole of solute is completely dissolved in enough solvent so that no further heat change takes place on adding more solvent (infinite dilution) at 298 K and 1 bar.

— *Standard enthalpy change of atomization*: The energy *absorbed* (i.e., endothermic) to form 1 mole of atoms in the gas phase from the element in the defined physical state at 298 K and 1 bar.

— *Standard enthalpy change of hydration*: The energy *evolved* when 1 mole of gaseous ions is dissolved in a large amount of water at 298 K and 1 bar.

— *Standard enthalpy change of neutralization*: The energy change when an amount of acid neutralizes a base to form 1 mole of water (in dilute aqueous solution) at 298 K and 1 bar.

(b) Cyclohexene reacts with hydrogen to form cyclohexane, C_6H_{12}, as follows:

$$C_6H_{10} + H_2 \rightarrow C_6H_{12}.$$

Calculate the ΔH for this reaction, given that the enthalpy change of formation of cyclohexene and cyclohexane are −36 kJ mol^{-1} and −156 kJ mol^{-1}, respectively.

Explanation:

$$6C(s) + 5H_2(g) \rightarrow C_6H_{10} \quad \Delta H_f(C_6H_{10}) = -36 \text{ kJ mol}^{-1}$$
$$6C(s) + 6H_2(g) \rightarrow C_6H_{12} \quad \Delta H_f(C_6H_{12}) = -156 \text{ kJ mol}^{-1}$$

By Hess' Law, ΔH of hydrogenation $= \Delta H_f(C_6H_{12}) - \Delta H_f(C_6H_{10}) = -120$ kJ mol^{-1}

Q Why is the ΔH of hydrogenation exothermic in nature?

A: This shows that the strength of 2 C–H bonds that are formed after hydrogenation is stronger than the pi bond in the C–C bond and the H–H bond in a H_2 molecule. If we use bond energy calculation:

$$\begin{aligned}\Delta H \text{ of hydrogenation} &= [\text{BE(C=C)} - \text{BE(C–C)}] + \text{BE(H–H)} - 2\text{BE(C–H)} \\ &= -124 \text{ kJ mol}^{-1}\end{aligned}$$

This is another testimony to the fact that if the bonds in the products are stronger than those in the reactants, the enthalpy change is exothermic!

(c) Benzene undergoes a similar reaction with hydrogen to form cyclohexane:

$$C_6H_6 + 3H_2 \rightarrow C_6H_{12}.$$

Assuming that benzene contains three C=C bonds of the type found in ethene, predict the value of ΔH for this reaction

Explanation:

ΔH of hydrogenation of benzene containing three C=C bonds of the type found in ethene = $3 \times (-124) = -372$ kJ mol^{-1}

(d) The actual value of ΔH for this reaction in part*(c)* is -210 kJ mol^{-1}. What can you deduce from this about the stability of the benzene ring? Use an energy level diagram to illustrate your answer.

Explanation:

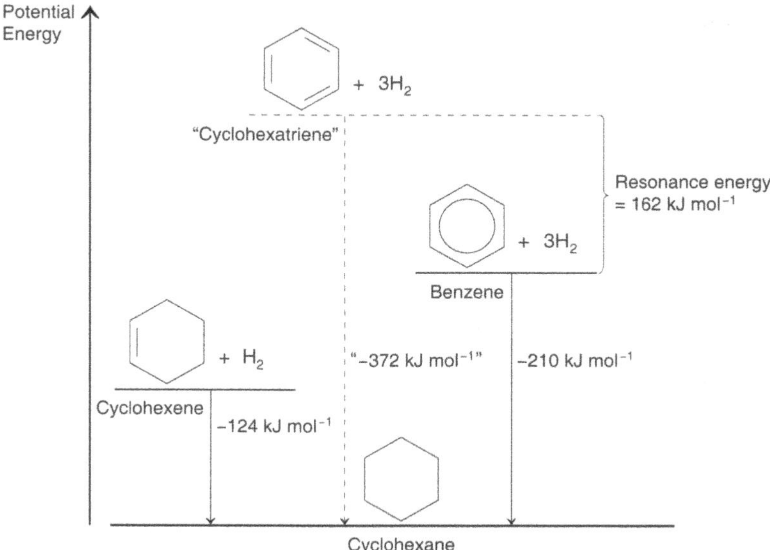

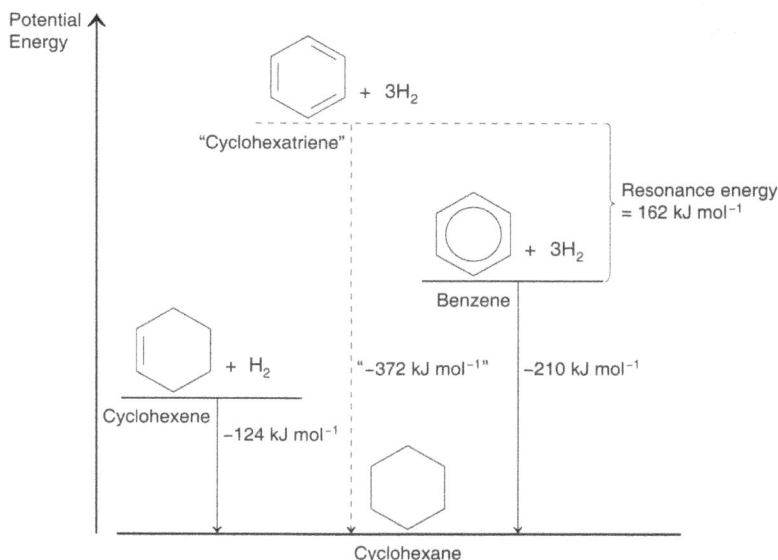

From the energy level diagram, since benzene releases less amount of energy than "cyclohexatriene," it must be at a lower energy level than "cyclohexatriene," which means benzene is more stable. The difference in energy is known as the resonance stabilization energy.

Q What is resonance?

A: Resonance refers to the delocalization of pi electrons through a network of overlapped *p* orbitals.

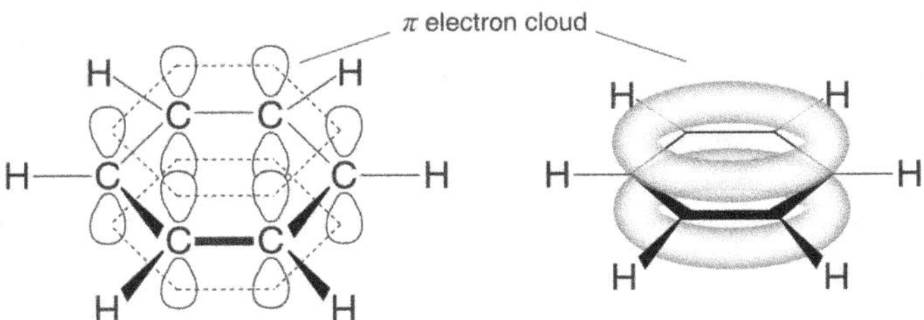

Resonance would bring about additional stability as when the electrons are allowed to move over a bigger space, there is minimization of inter-electronic repulsion.

5. (a) State Hess's Law.

Explanation:

Hess' Law states that the enthalpy change for a chemical process is only dependent on the initial and final states but independent on the pathway taken.

(b) (i) Using the following data:

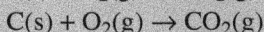

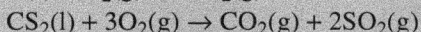

$S(s) + O_2(g) \rightarrow SO_2(g)$ $\Delta H^\theta = -297\ \text{kJ mol}^{-1}$
$C(s) + O_2(g) \rightarrow CO_2(g)$ $\Delta H^\theta = -393\ \text{kJ mol}^{-1}$
$CS_2(l) + 3O_2(g) \rightarrow CO_2(g) + 2SO_2(g)$ $\Delta H^\theta = -297\ \text{kJ mol}^{-1}$,
calculate the enthalpy change for the reaction between carbon and sulfur to form carbon disulfide:

$$C(s) + 2S(s) \rightarrow CS_2(l).$$

Explanation:

$S(s) + O_2(g) \rightarrow SO_2(g)$ $\quad \Delta H_1^\theta = -297\ \text{kJ mol}^{-1}$
$C(s) + O_2(g) \rightarrow CO_2(g)$ $\quad \Delta H_2^\theta = -393\ \text{kJ mol}^{-1}$
$CS_2(l) + 3O_2(g) \rightarrow CO_2(g) + 2SO_2(g)$ $\quad \Delta H_3^\theta = -297\ \text{kJ mol}^{-1}$

$$\Delta H_r = 2\Delta H_1^\theta + \Delta H_2^\theta - \Delta H_3^\theta = -690\ \text{kJ mol}^{-1}$$

(ii) Is the enthalpy change you have calculated equal to the standard enthalpy change of formation of carbon disulfide? Explain your answer.

Explanation:

The definition of ΔH_f^θ is: the energy change when 1 mole of a pure compound in a specified state is formed from its constituent elements in their standard states, at 298 K and 1 bar. Thus, according to the equation, we have 1 mole of CS_2 in the liquid state formed from its constituent elements in their standard state. Yes, it is equivalent!

(iii) Metal sulfide ores are usually roasted in air to form the oxide before reduction to the metal with carbon. Explain this practice with reference to your answer in part *(b)(i)*.

Explanation:

When the metal sulfide ores are roasted in air, the sulfide will be converted to SO_2, which can be captured and converted to sulfuric acid. In addition, if the sulfide ores are not converted to the oxide, and then when carbon is added to reduce the sulfide ores, the carbon will react readily with the sulfide to form CS_2. This is a waste of resources.

(c) In a typical experiment using an acid and an alkali of equal volumes, the following results were obtained:

Concentration of sulfuric(VI) acid = 1.00 mol dm^{-3}
Concentration of sodium hydroxide = 2.00 mol dm^{-3}
Volume of sulfuric(VI) acid used = 25.0 cm^3
Initial temperature of sulfuric(VI)acid = 21.0 °C
Initial temperature of sodium hydroxide = 23.0 °C
Highest temperature reached after mixing = 35.6 °C
Specific heat capacity of water = 4.2 J g^{-1} K^{-1}

(i) Calculate the molar enthalpy change of neutralization of sulfuric(VI) acid with sodium hydroxide from these results. Indicate the assumptions you have made.

Explanation:

Amount of H_2SO_4 used = 25/1000 × 1.00 = 0.025 mol
Amount of NaOH used = 25/1000 × 2.00 = 0.050 mol
Total volume of solution = 50.0 cm^3
Weighted average initial temperature of mixture = $\frac{25\times21+25\times23}{50}$ = 22.0 °C
Hence, ΔT = 35.6 − 22.0 = 13.6 °C
Heat change = $mc\Delta T$ = 50.0 × 4.2 × 13.6 = −2856 J

$$\frac{1}{2}\,H_2SO_4(aq) + NaOH(aq) \rightarrow \frac{1}{2}\,Na_2SO_4(aq) + H_2O(l)$$

Amount of H_2O produced = Amount of NaOH used = 0.050 mol
Enthalpy change of neutralization = $\frac{\text{Heat change}}{\text{Amount of water produced}}$ = $\frac{-2856}{0.050}$ = −57.1 kJ mol^{-1}

Do you know?

— You need to calculate the weighted average temperature of the mixture when you mix two or more solutions of different initial temperatures together.

(Continued)

(Continued)

- The so-called mass of the mixture is actually the volume of the mixture. This is because the heat energy that is evolved is "shouldered" by the water molecules, which is present in large excess.
- The enthalpy change of neutralization is always calculated as per mole of water molecules formed. So, importantly, the amount of water formed is affected by whether the acid or base is the limiting reagent.
- If the neutralization equation is $H_2SO_4(aq) + 2NaOH(aq) \rightarrow Na_2SO_4(aq) + 2H_2O(l)$ instead, then the enthalpy change of neutralization is -114.2 kJ mol^{-1}. This brings us to the importance of defining an enthalpy change with an appropriate equation in accordance to the theoretical definition of the enthalpy term. The 'per mole' in the kJ mol^{-1} unit is actually per mole of the chemical equation that is being defined!
- The enthalpy change of neutralization between a strong acid and a strong base is about -57.1 kJ mol^{-1}. But if a weak acid or weak base or both are weak are used, then the enthalpy change of neutralization can be either more exothermic or less exothermic than -57.1 kJ mol^{-1}.

Q What is a strong acid or a strong base?

A: An acid or a base is considered strong if it fully dissociates in water. For example, HCl, HBr, HI, H_2SO_4, HNO_3 NaOH, KOH, $Ba(OH)_2$, etc., are all considered strong acids or bases. Weak acids or bases do not dissociate or ionize fully in water. For example, NH_3, organic amines, CO_3^{2-}, carboxylic acids, phenols, HF, HCN, etc., are all weak acids or bases.

Q So, why would the enthalpy change of neutralization of a weak acid or base not be equal to -57.1 kJ mol^{-1}?

A: As the weak acid or base does not fully dissociate in water, part of the heat energy that is given off during neutralization would be diverted to help further dissociate or ionize the weak acid or base. This would hence make the enthalpy change of neutralization to be less negative or less exothermic than -57.1 kJ mol^{-1}.

Q So, we can actually use the measurement of the enthalpy change of neutralization to differentiate a strong acid from a weak one?

A: Absolutely right! It also works for differentiating a strong base from a weak one.

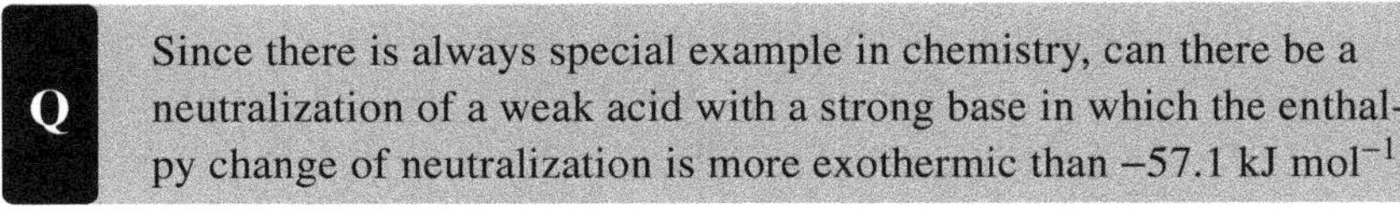

Q Since there is always special example in chemistry, can there be a neutralization of a weak acid with a strong base in which the enthalpy change of neutralization is more exothermic than -57.1 kJ mol^{-1}?

A: Of course! The enthalpy change of neutralization of HF with NaOH is actually more exothermic than -57.1 kJ mol^{-1}, although HF is a weak acid. It is in fact about -68.0 kJ mol^{-1}. This is because the F^- ion that is generated is small and due to its high charge density, it forms very strong ion–dipole interaction with the water molecules. Hence, this releases much heat energy in the form of enthalpy change of hydration.

(ii) Calculate the percentage error in your result assuming that temperatures were accurate to ±0.1°C.

Explanation:

Error in $\Delta T = \pm(0.1 + 0.1) = \pm 0.2^\circ C$ (errors add up!)
Hence, percentage error due to temperature measurement $= \dfrac{\pm 0.2 \times 100}{22} = \pm 0.1\%$

(iii) Account for the fact that a similar experiment using hydrochloric acid gave an identical value for the molar enthalpy of neutralization, while an experiment with methanoic acid gave a value of -55.0 kJ mol^{-1}.

Explanation:

Methanoic acid, HCOOH, is a weak acid that does not fully dissociate in water. As such, part of the heat energy that is given off during neutralization would be diverted to help further dissociate the weak acid, causing the enthalpy change of neutralization to be less exothermic than -57.1 kJ mol^{-1}.

(d) Describe how you would carry out an experiment to determine the molar enthalpy change of sulfuric(VI) acid with sodium hydroxide. Your account should mention the measurements which you would make, and the reasons behind your choice of apparatus.

Explanation:

The following set-up is used:

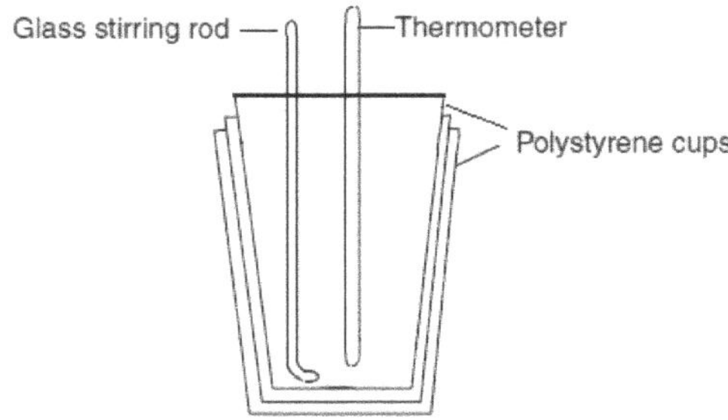

(1) A measured volume of NaOH(aq) of known concentration is placed in an insulated polystyrene cup. This ensures that there is minimal heat lost to the surrounding.

(2) The initial temperature (T_i) of the solution is recorded before the start of the reaction. At the same time, the initial temperature of a known volume of H_2SO_4(aq) with known concentration is also measured. By measuring the initial temperatures of both solutions, it ensures an accurate measurement of the actual initial temperature of the reaction mixture.

(3) At a suitable time interval, the measured volume of H_2SO_4(aq) is added to the cup. Continuous stirring is performed so as to ensure that all the heat energy is quickly dispersed throughout the solution. In addition, this also ensures that all the reactants "react at the same time" and all the heat energy is released at "one go."

(4) The highest temperature of the resultant solution is recorded in the shortest time so that there is minimal heat lost to the surroundings.

No matter what, there is still heat lost to the surroundings. Is there a way to circumvent it?

A: Yes, of course. What you can do is while you are measuring the initial temperature of the solution that sits in the polystyrene cup, start a stopwatch. Measure the temperature at regular time interval, say every 30 sec. Then, at an appropriate time, let's say at the 3 minute point, pour in the other solution. Stir quickly to mix well and at the same time monitor the temperature at regular time intervals. Then, plot a graph of the measured temperature versus time. Do the following extrapolation and find the maximum temperature at the point of mixing, which is at 3 min. The difference from the initial temperature to the maximum temperature at the 3 min point is the theoretical change of temperature without any heat lost to the surroundings.

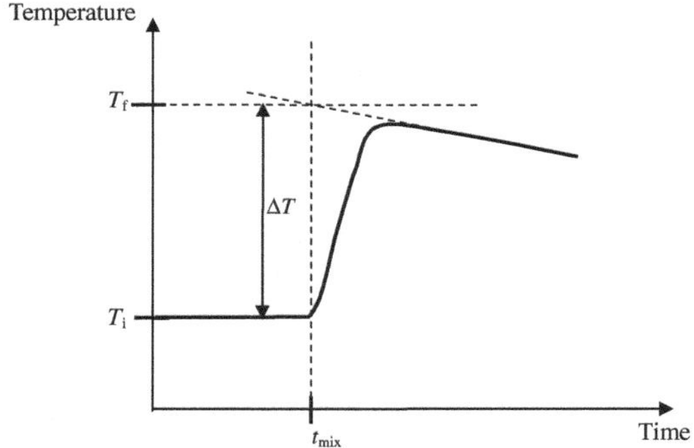

CHAPTER 5

REACTION KINETICS

1. Acid rain contains dissolved sulfuric(VI) acid which has been formed by the atmospheric oxidation of sulfur dioxide by other pollutants produced from combustion. The mechanism involved, together with their rate constants (k), are shown below. (All these rate constants have units of $mol^{-1}\ dm^3 min^{-1}$.)

$$HO\cdot + SO_2 \rightarrow HOSO_2\cdot \qquad k = 7.2 \times 10^{10}, \qquad (1)$$
$$HOSO_2\cdot + O_2 \rightarrow HO_2\cdot + SO_3 \qquad k = 1.44 \times 10^{10}, \qquad (2)$$
$$SO_3 + H_2O \rightarrow H_2SO_4 \qquad k = 2.16 \times 10^{8}, \text{ and} \qquad (3)$$
$$HO_2\cdot + NO\cdot \rightarrow HO\cdot + NO_2\cdot \qquad k = 3.0 \times 10^{11}. \qquad (4)$$

(a) Suggest why the oxidation of SO_2 to H_2SO_4 occurs more rapidly in air which contains other gaseous pollutants produced from combustion.

Explanation:

The other gaseous pollutants, such as NO, can help in step (4) in the above mechanism to generate $HO\cdot$, which can catalyze the conversion of SO_2 to $HOSO_2\cdot$ in step (1).

Q What is a mechanism?

A: A mechanism is a sequence of steps that shows how the reactants participate and in what amounts, leading to the final products. Each step of a mechanism is an elementary step, meaning, it cannot be broken down into any other steps that are simpler than itself. They are irreducible! It is just like a prime number, which can only be divisible by one and itself. And among these steps, there is one step that is the slowest step among all. It is known as the 'slow step' or 'rate-determining step,' in short, r.d.s.

Do you know?

— A catalyst is a reactant that speeds up the rate of a reaction but remains unchanged at the end of the reaction. As such, a catalyzed pathway has a lower activation energy than an uncatalyzed pathway.

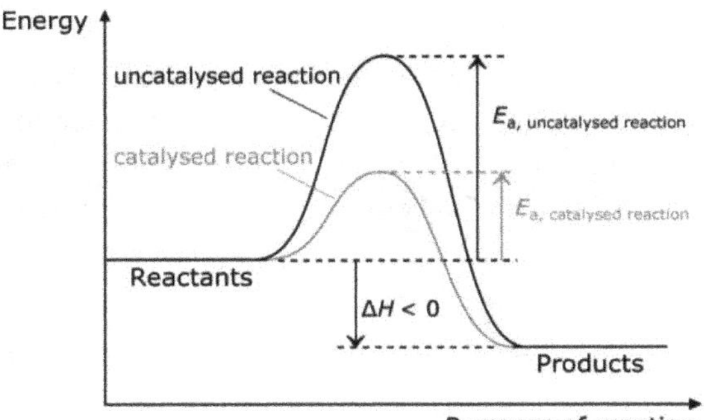

The activation energies for both the forward and backward reactions are being decreased by the same amount! In addition, the enthalpy change of reaction is not altered when a catalyst is introduced.

— As a result of the lowered activation energy, there are more particles with kinetic energy greater than the activation energy, hence the frequency of successive collisions increases.

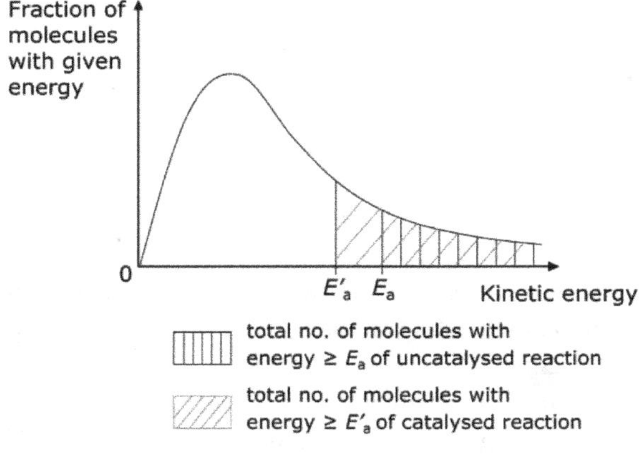

If a catalyst is actually a reactant, and if we increase the concentration of the catalyst, then the rate of the reaction should be increased, right?

A: Yes, of course. If it is a homogeneous catalysis, in which both the reactants and catalyst are in the same phase, then increasing the concentration of catalyst would increase the frequency of successful collisions of the catalyst with the reactants. This would lead to a higher reaction rate. And if it is a heterogeneous catalysis, in which both the reactants and the catalyst are of different phases, then an increase in the surface area of the solid catalyst would increase the number of active sites whereby the reactants can adsorb. This would thus increase the rate of reaction.

Q Why is there an activation energy needed?

A: Particles need to "touch" each other in order to react. But when they approach each other, there would be greater inter-electronic repulsion between their electron clouds. Hence, the particles need to have sufficient kinetic energy to let their electron cloud inter-penetrate each other, so as to facilitate electron movement in the chemical reaction. In addition, the particles need to be in the correct orientation and bonds need to be broken and rearranged. All these processes need energy and are in the form of activation energy.

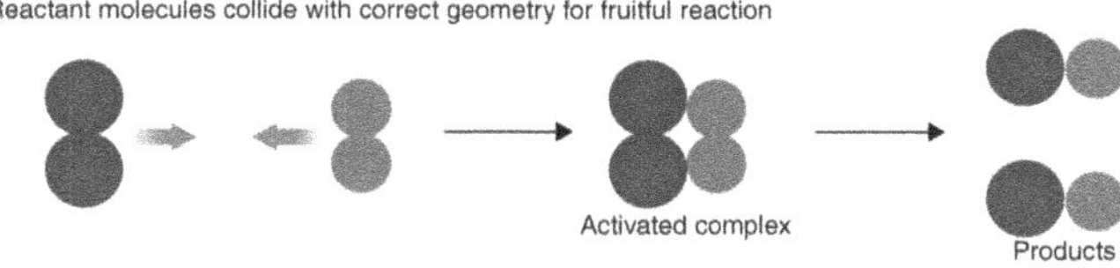
Reactant molecules collide with correct geometry for fruitful reaction
Activated complex
Products

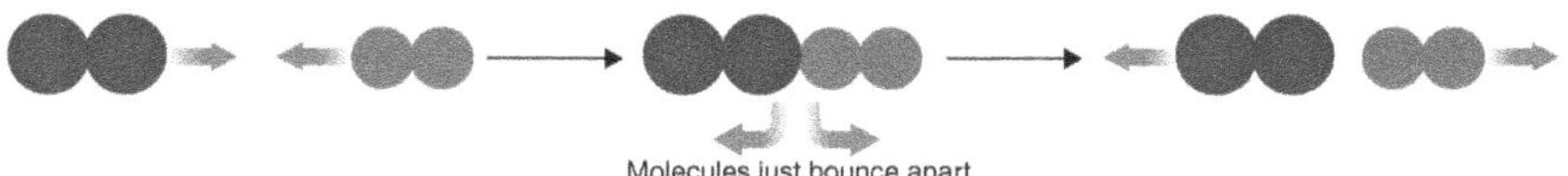
Wrong collision geometry leads to unfruitful reaction
Molecules just bounce apart

So, activation energy literally is the energy needed to activate a reaction. And they are not the same in an exothermic or endothermic reaction.

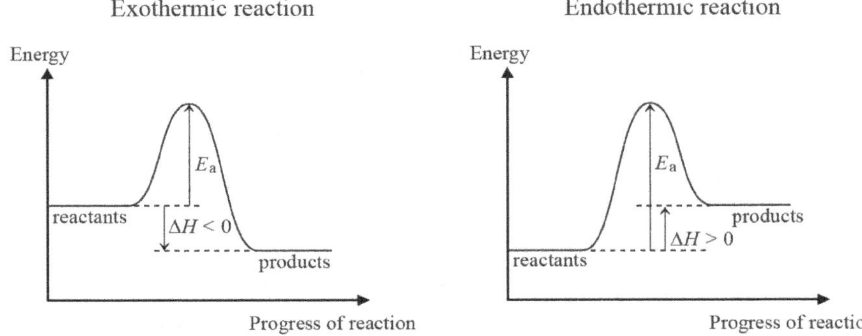

Q So, does that mean that activation energy is also needed in a precipitation reaction involving free ions?

A: Yes, of course. And do not forget, the aqueous ions are heavily hydrated and when two ions meet, it is equivalent to the meeting of two giant electron clouds. So, there is bound to be inter-electronic repulsion. In addition, energy is also needed to remove these water molecules by breaking the ion–dipole interactions. Hence, there would be an additional amount of activation energy involved too.

(b) Explain the meaning of the term *rate constant.*

Explanation:

The rate constant, k, is a proportionality constant in the rate equation of the reaction.

What is a rate equation?

A: A rate equation, which is also known as the rate law, is a mathematical equation that shows how the rate of reaction is dependent on the concentrations of the reactants; it relates the rate of the reaction to the concentrations of reactants raised to the appropriate power. For example, rate = $k[A]^m[B]^n$.

Q What are m and n?

A: They are known as the order of the reaction: (i) the order of reaction with respect to a reactant is the power to which the concentration of that reactant is being raised to in the rate law; and (2) the overall order of reaction is the sum of the powers to which the concentrations of the reactants are being raised to in the rate law. So, if the reaction is a non-zero order with respect to a particular reactant, this would mean that increasing the concentration would increase the rate of reaction.

Low concentration

Higher concentration

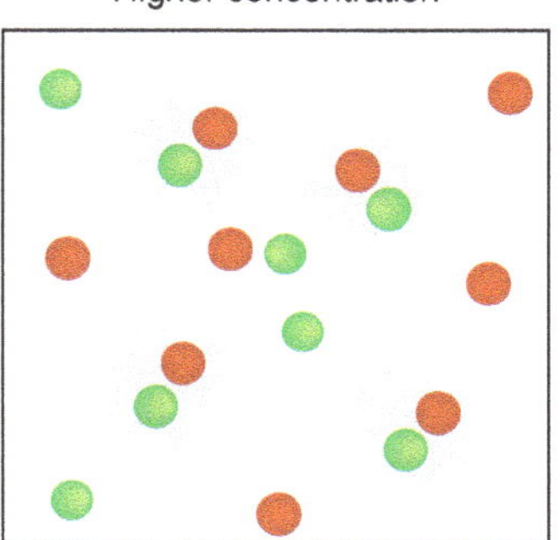

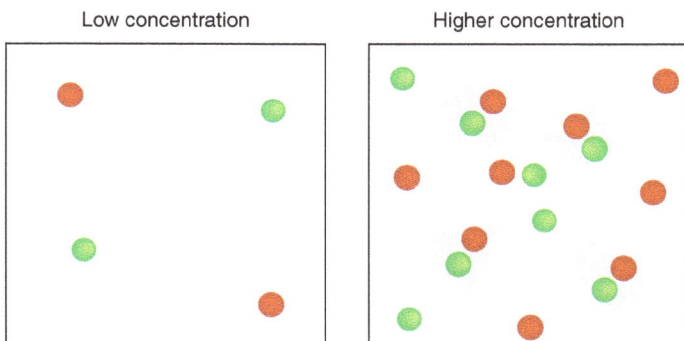

Q Can the order of reaction be a non-integral number or even a negative number?

A: Yes, of course. A negative order would mean that when we increase the concentration, it actually decreases the rate. This would mean that the reactant is in fact some form of an inhibitor as it inhibits the reaction. A non-integral order of reaction is too complicated to be explained here, so we would not delve into it.

 But is there any physical meaning behind the order of a reaction?

A: In order to understand this, we need to first understand what a rate law is all about. First, a rate law is obtained experimentally and it measures the rate of the reaction. So, the rate equation is actually a measurement of how fast the rate of the r.d.s. is in the whole mechanism. With this concept, it is not difficult to understand that the order of reaction in the rate law depicts the amount of that particular reactant that appears in the r.d.s. In short, the order of reaction indicates the molecularity of the reactant in the r.d.s.

Do you know?

— The common rate constant expression that we use is the Arrhenius rate constant, $k = \text{A}\exp\left(-\frac{\text{E}_\text{a}}{\text{RT}}\right)$. Hidden in this expression are two variables that affect the rate of a reaction, (i) the activation energy, E_a, which we have discussed, and (ii) the temperature factor, T.

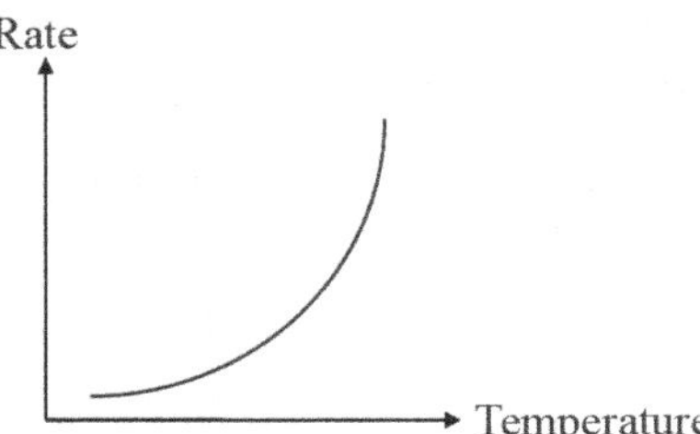

— When we increase the temperature of the system, the rate of reaction usually increases because at a higher temperature, (i) we now have more particles with kinetic energy greater than the activation energy; this leads to a higher frequency of effective collisions, and (ii) since the particles have more kinetic energy, they move faster, hence the higher frequency of collisions would lead to more effective collisions. The temperature effect can be represented using the Maxwell–Boltzmann distribution plot below:

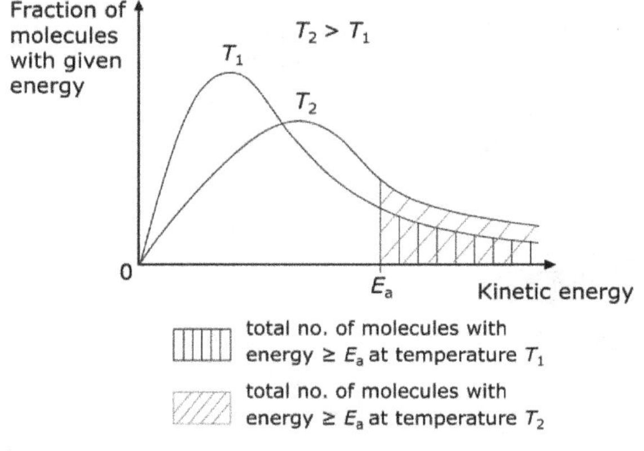

Why did you say that "when we increase the temperature of the system, the rate of reaction usually increases because it is at a higher temperature?" Do you mean to say that there are reactions where when you increase the temperature, the rate of reaction actually decreases?

A: Yes. Take for instance, an enzyme-catalyzed reaction. At a temperature that is beyond the optimal working temperature of the enzyme, the enzyme may denature. Hence, it would lose its activity, therefore the rate of reaction would decrease.

So, the rate constant k translates both the physical effects brought about by the activation energy and the temperature factors together into a mathematical relationship?

A: Well said! Indeed, scientists were able to transform the physical effect into a mathematical formula and hence make it more manageable and neater.

(c) Laboratory experiments have been done to investigate the reaction between SO_2 and HO· (Eq. (1)) under conditions similar to those present in polluted air. The data in the table below show how the concentration of HO· changes with time, for an experiment in which the sulfur dioxide concentration is 5×10^{-5}mol dm^{-3}.

Concentration of HO·/10^{-8}mol dm^{-3}	2.0	1.5	1.1	0.8	0.6	0.4
Time (t)/10^{-8}min	0	8.33	16.67	25.00	33.33	41.67

(i) Plot these data on a piece of graph paper.

Explanation:

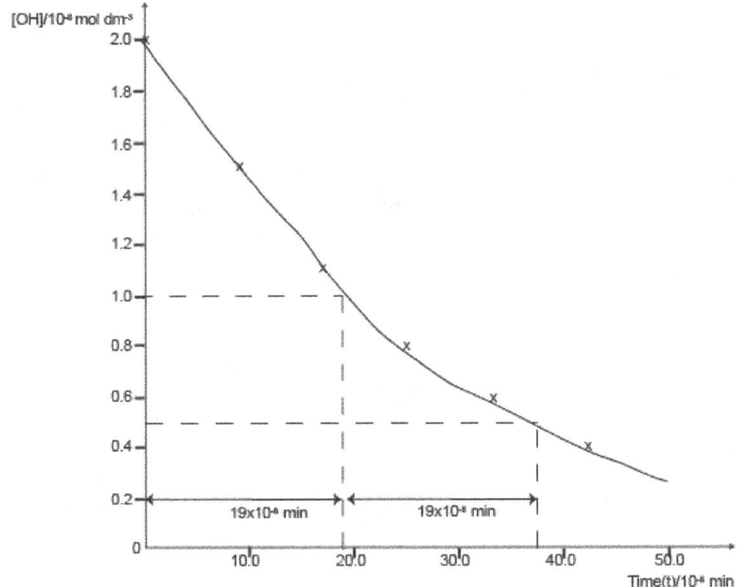

(ii) Using the half-life method, show that the reaction is first order with respect to HO·.

Explanation:

Since the two half-lives are constant, $t_{½} = 19 \times 10^{-8}$ min, the reaction is first order with respect to [HO·].

Q Why is the half-life for a first order reaction a constant value?

A: A first-order rate equation looks like, this: rate = k[A]. After integration, you would get a equation which shows how [A] varies with time t:

$[A] = [A]_o e^{-kt}$ where $[A]_o$ is the initial concentration of reactant A.

Half-life is the time taken ($t_{½}$) for the [A] to be $\frac{[A]_0}{2}$:

$$\text{i.e.}\quad \frac{[A]_0}{2} = [A]_0 e^{-kt} \Rightarrow \tfrac{1}{2} = e^{-kt} = (2)^{-1}$$
$$\Rightarrow \ln 2 = kt_{1/2}$$
$$\Rightarrow t_{1/2} = \frac{\ln 2}{k}$$

So, in the $t_{1/2}$ equation, [A]'s term does not appear, hence the $t_{1/2}$ is independent of the starting concentration of the reactant.

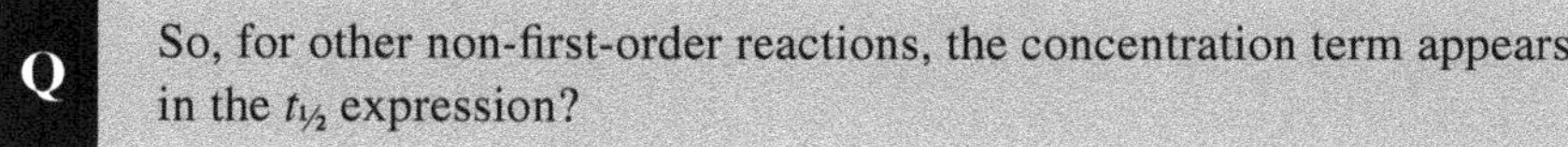

Q So, for other non-first-order reactions, the concentration term appears in the $t_{1/2}$ expression?

A: Yes, indeed. For a zero-order reaction, the $t_{1/2} = \frac{[A]_0}{2k}$ while for a second-order reaction, the $t_{1/2} = \frac{1}{k[A]_0}$. Hence, the half-lives of zero-order and second-order reactions are dependent on the starting concentration of the reactant.

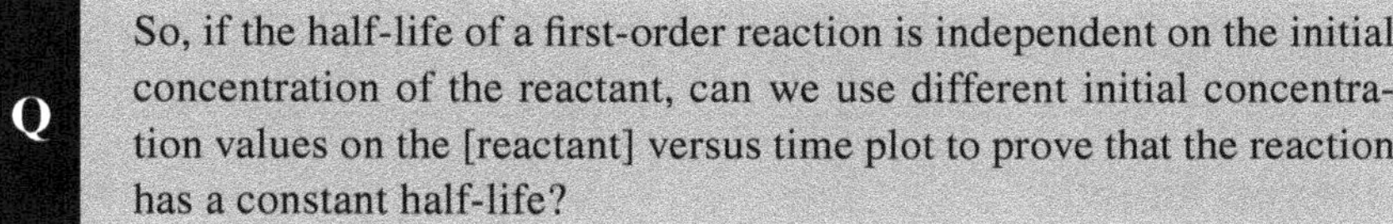

Q So, if the half-life of a first-order reaction is independent on the initial concentration of the reactant, can we use different initial concentration values on the [reactant] versus time plot to prove that the reaction has a constant half-life?

A: Yes, indeed. For the graph above, you can find the time taken for [HO·] to go from 1.2×10^{-8}mol dm^{-3} to 0.6×10^{-8}mol dm^{-3}. It would still be around 19×10^{-8} min. Try it!

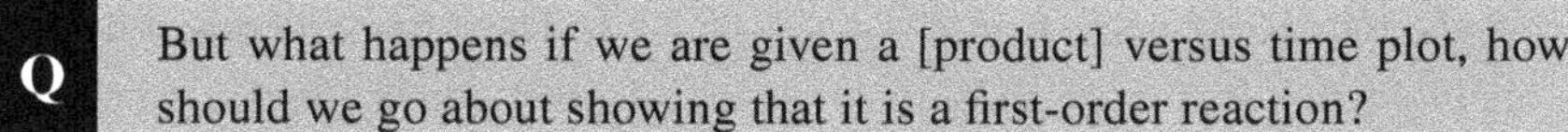

Q But what happens if we are given a [product] versus time plot, how should we go about showing that it is a first-order reaction?

A: At the point $t = 0$, the [reactant] is the maximum while [product] = 0. So, for the first $t_{1/2}$, when the [reactant] is halved, the [product] increased to half of its maximum, i.e., to ½ C_0. Now, the [reactant] is left with half of its original amount. The next half-life would be the time taken for the [reactant] to go from half of its original concentration to ¾ of its original concentration. This would be equivalent to the time needed for the product to go from half of its maximum concentration to ¾ of its maximum concentration.

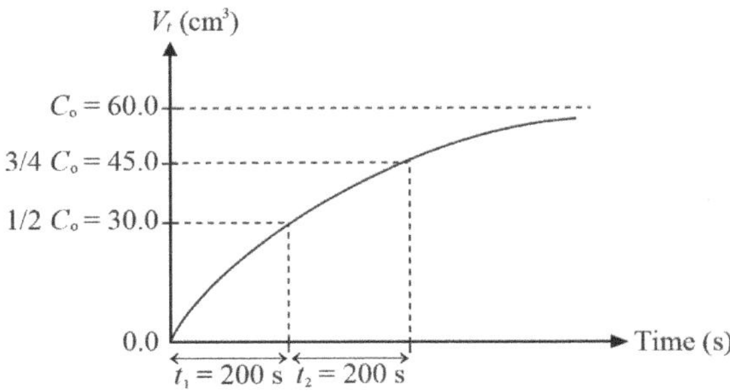

Q For the [product] versus time plot, if we do not have the value for the maximum [product], can we just arbitrarily take any point on the graph and treat it as the maximum point? And from there use it to determine whether half-life is a constant?

A: You can't! What you can do is to try to estimate the asymptotic maximum [product].

Do you know?

— In order to prove that the reaction is first-order, one can plot [reactant] versus time, and from this plot show that the half-lives are constant. Other than this, one can also calculate the rate of the reaction at different concentration points on the graph by drawing different tangents. The gradient of each tangent at a particular concentration point gives the rate since gradient = $\frac{\text{chane of concentration}}{\text{change of time}}$. Then, one can proceed to plot a graph of rate versus [reactant]. If the outcome is a linear straight line, the reaction is a first-order one since rate = k[reactant].

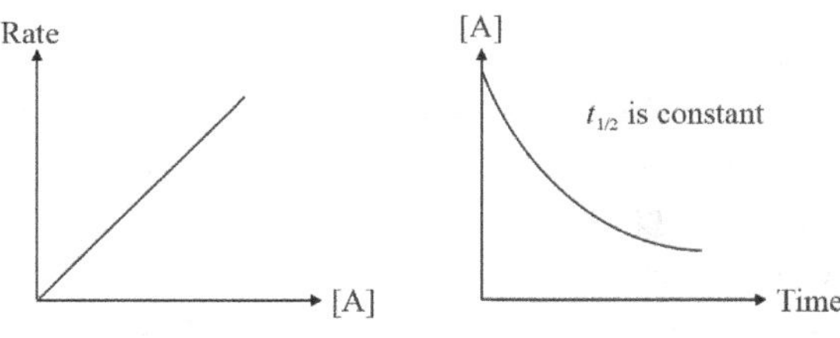

— The useful thing about plotting rate versus [reactant] is that if you find that it is not a straight line but instead a curve, you can then proceed to

(Continued)

(*Continued*)

plot a rate versus [reactant]^2. And if this turn out to be a straight line, then it is a second-order reaction.

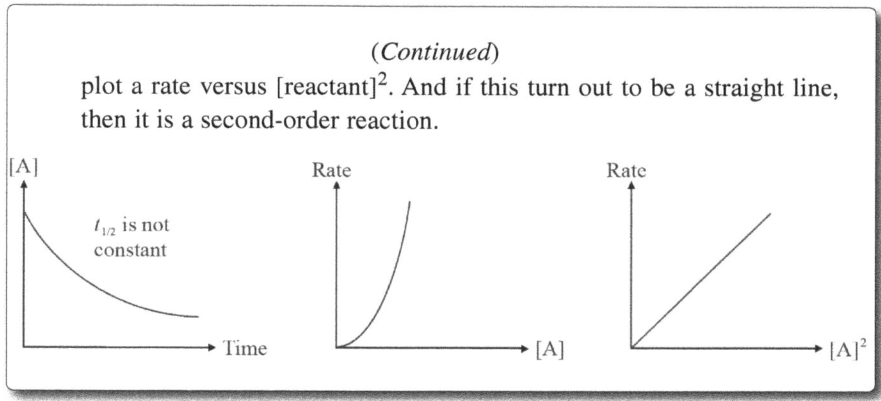

Q How about showing that it is a zero-order reaction?

A: Simple! If you plot the [reactant] versus time graph and find it to be a decreasing linear straight line it is zero order. Or if you plot the rate versus [reactant] graph and find it to be a horizontal straight line, that is it! This is because mathematically, rate = k, a constant for a zero-order reaction.

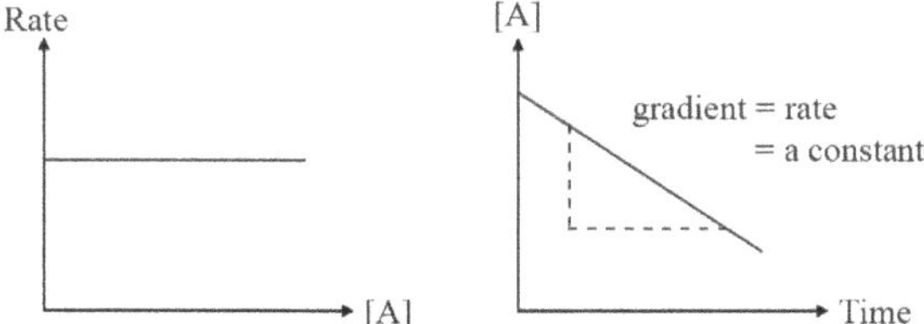

(iii) Explain what is meant by the statement: "*Reaction is first order with respect to HO·.*"

Explanation:

This means that the rate equation is rate = k[HO·]. The power to which the [HO·] is being raised is one, and this is the order of reaction with respect to [HO·]. In addition, if we double the [HO·], the rate would also be doubled.

2. The reaction between bromate(V) and bromide ions in acid solution forming bromine follows the equation:

$$BrO_3^-(aq) + 5Br^-(aq) + 6H^+(aq) \rightarrow 3Br_2(aq) + 3H_2O(l).$$

The rate of this reaction can be studied using the initial rate method. A small fixed amount of phenol and an acid–base indicator are added to color the solution. The phenol undergoes a rapid initial reaction with the bromine, resulting in a clear solution. When this reaction is completed, the next portion of bromine bleaches the indicator. Finally, a white precipitate is slowly formed and the solution eventually turns orange as excess bromine is produced. The time, t, taken until the solution is colorless can be used as a measure of the rate of reaction.
A study of the kinetics of the above reaction reveals that it is second order with respect to H^+ and first order with respect to BrO_3^- and Br^-.

(a) What is the overall order of the reaction?

Explanation:

The rate equation is rate = $k[H^+]^2[BrO_3^-][Br^-]$, hence overall order of the reaction is 4.

Q What is an initial rate method?

A: The initial rate method refers to a method commonly used in kinetics study in which the average rate of reacting a fixed concentration of a reactant or forming a fixed concentration of product is actually measured. In this example, how fast Br_2 is produced would depend on how fast the various reactants react. And the time taken for a fixed amount of Br_2 to be produced is the same time taken for fixed amount of phenol to react with the Br_2. It is like a 100-m race. All the runners have to cover the same distance and the one that takes the shortest time to cover this distance as the rest of the runners, must be the fastest runner! Thus, mathematically, initial rate $\alpha \frac{\Delta c}{\Delta t}$ and if Δc is the same for all experimental sets, then initial rate $\alpha \frac{1}{\Delta t}$.

Why can't we just measure the time taken for the orange Br_2 to appear without adding phenol into the reaction mixture?

A: The moment all the reactants are mixed, the orange Br_2 appear instantaneously. How does the human reaction time respond to the differences? We need to add something to "stall' the appearance of the orange Br_2, to allow us more time to response as accurately as possible.

Q Why did you say that the initial rate is an average rate?

A: The actual initial rate refers to the gradient of the tangent drawn at the point when $t = 0$ on the [reactant] versus time plot.

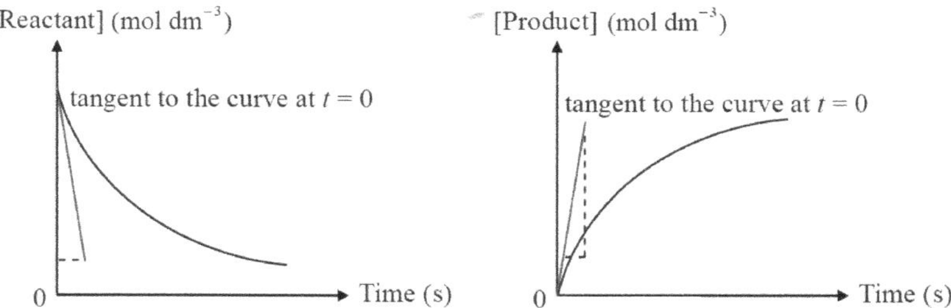

At this point, all the concentration of the reactants are known. But this point can only be obtained if you plot the [reactant] versus time graph. When we measure the so-called initial rate using the initial rate method, a certain amount of time would already have lapsed. Hence, what we are really measuring is the average rate.

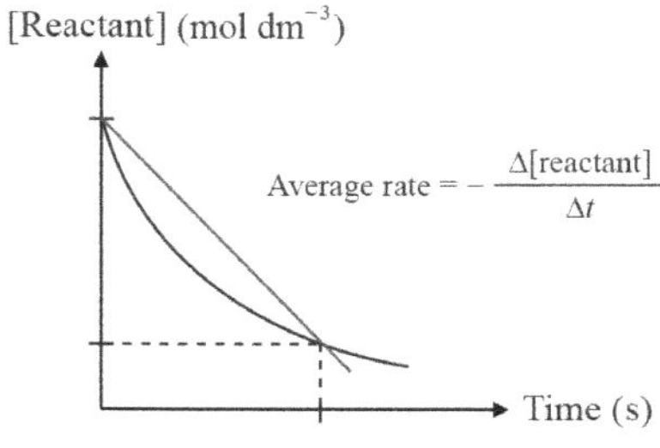

Q So, does it mean that the Δc or Δt measured cannot be too great?

A: Yes! Too great a Δc or Δt measured would pose a problem because the concentration of the reactants at the point of measurement may differ too greatly from the initial values. This would invite inaccuracy in our assessment of the order of reaction. But neither can they be too small to be measured, as the smaller the measurement, the greater the amount of error incurred in the measurement.

(b) Using the collision theory, explain why this reaction is unlikely to occur in a single step.

Explanation:

According to the collision theory, the particles need to collide with one another in order to react. In this case here, the collision involves four particles at one go, which is statistically improbable. Hence, the reaction is unlikely to occur in a single step.

But it was mentioned that the order of reaction reflects the number of particles that take part in the slow step. So, if the slow step for the above reaction cannot involve the collision of four particles, how does the slow step looks like?

A: Statistically, collision involving three particles is already quite improbable, not to mention if the reaction involves four particles. Usually for an overall order of reaction that is above 2, there should be other mechanistic steps before the slow steps. And because the slow step is the r.d.s., this would mean that all other steps before this step would reach an equilibrium state. Take for instance the mechanism for the above reaction, which has a pre-equilibrium mechanism as follows:

$BrO_3^- + H^+ \rightleftharpoons HBrO_3$ $\quad k_{f1}[BrO_3^-][H^+] = k_{b1}[HBrO_3]$

$HBrO_3 + H^+ \rightleftharpoons H_2BrO_3^+$ $\quad k_{f2}[HBrO_3][H^+] = k_{b2}[H_2BrO_3^+]$

$H_2BrO_3^+ + Br^- \rightarrow (Br\text{–}BrO_2) + H_2O$ $\quad \text{rate} = k_{f3}[H_2BrO_3^+][Br^-]$

$(Br\text{–}BrO_2) + 4H^+ + 4Br^- \rightarrow 3Br_2 + 3H_2O$

$[H_2BrO_3^+] = \frac{k_{f2}}{k_{b2}}[HBrO_3][H^+] = \frac{k_{f2}}{k_{b2}}[H^+].\frac{k_{f1}}{k_{b1}}[BrO_3^-][H^+] = \frac{k_{f2}}{k_{b2}}\frac{k_{f1}}{k_{b1}}[BrO_3^-][H^+]^2$

Substituting into rate $= k_{f3}[H_2BrO_3^+][Br^-]$

$= k_{f3}\frac{k_{f2}}{k_{b2}}\frac{k_{f1}}{k_{b1}}[BrO_3^-][H^+]^2[Br^-]$

$= k[BrO_3^-][H^+]^2[Br^-]$,

which is similar to the experimental rate law.

Q How do we know whether a mechanism is a possible one?

A: Two criteria need to be fulfilled: (1) the sum of the mechanistic steps must be equal to the overall reaction; and (2) the rate law for the slow step must be equivalent to the experimental rate equation!

(c) By what factor will the rate of reaction be reduced if an equal volume of water is added to a sample of the reacting mixture?

Explanation:

By adding an equal volume of water, the concentration of each of the reactants is decreased by half. Hence, the rate would be decreased by a factor of $(1/2)^4 = 1/16$.

Do you know?

— Adding excess water is a way in kinetics study to slow down a reaction. It is known as 'quenching.' Increased volume and decreased concentration, hence the reactants are farther apart and according to the collision theory, less probable to react.

— Adding cold water would be a better way to quench the reaction. This is because when cold water is added, the reacting temperature decreases. Hence, the rate of reaction also decreases.

(d) Calculate the rate of bromine formation if the initial concentrations of bromate(V), bromide and H^+ are 0.04, 0.2, and 0.1mol dm^{-3}, respectively, and the rate constant is 2×10^{-2} dm^3mol^{-3} min^{-1}.

Explanation:

Rate = $k[BrO_3^-][H^+]^2[Br^-] = (2 \times 10^{-2})(0.04)(0.1)^2(0.2) = 1.6 \times 10^{-6}$mol dm^{-3} min^{-1}.

(e) Suggest why the orders with respect to Br^- and H^+ are not equal to the coefficients in the stoichiometric equation for the reaction.

Explanation:

A reaction can consist of a series of elementary steps in which the sum of these steps is equal to the overall reaction equation. The order of a reaction reflects the number of particles participating in the slow step, which naturally explains why it may not be equivalent to the coefficients in the overall reaction equation.

So, can there be a reaction where the order of reaction is equivalent to the coefficient in the overall reaction equation?

A: Yes, of course!

(f) Suggest why time *t* is inversely proportional to the rate constant, *k*, given the conditions in part (*d*).

Explanation:

Time *t* is the time taken for a fixed amount of phenol to react with a fixed amount of Br_2. Hence, the rate of reaction is inversely proportional to *t*, i.e., rate $\alpha \frac{1}{t}$.

But rate = $k[BrO_3^-][H^+]^2[Br^-] \propto \frac{1}{t}$. Therefore, $k \propto \frac{1}{t}$.

(g) The activation energy, E_a, for the reaction is related to the time, t, by the equation (where T is temperature in Kelvin):

$$\ln t = \text{constant} + E_a/RT.$$

Use the data below to calculate a value for E_a.

T/°C	26	49
t/s	148	24

Explanation:

With T = 299 K, t = 148 s,

$$\ln 148 = \text{constant} + E_a/(8.314 \times 299) = \text{constant} + 4.02 \times 10^{-4}\ E_a$$

With T = 322 K, t = 24 s,

$$\ln 24 = \text{constant} + E_a/(8.314 \times 322) = \text{constant} + 3.74 \times 10^{-4}\ E_a$$

Solving it simultaneously, E_a = 64969.9 J mol^{-1} = 64.97 kJ mol^{-1}.

Do you know?

— This is how scientists can experimentally determine the activation energy of a reaction. It is derived from the Arrhenius rate constant, k:

$$k = A \exp\left(-\frac{E_a}{RT}\right) \Rightarrow \ln k = \ln A - \frac{E_a}{RT}$$

$$\text{But since } k \propto \frac{1}{t}, \Rightarrow \ln\left(\frac{1}{t}\right) = \ln A - \frac{E_a}{RT}$$

$$\Rightarrow \ln t = \ln\left(\frac{1}{A}\right) + \frac{E_a}{RT} = \text{constant} + \frac{E_a}{RT}.$$

Q What is the meaning of 'A' in the Arrhenius rate constant, $k = A\exp\left(-\frac{E_a}{RT}\right)$?

A: The constant 'A' is known as the Arrhenius constant or other names such as 'frequency factor,' 'pre-exponential factor,' or 'steric factor.' It is a measure of the proportion of molecules (frequency factor) that collide with enough energy to react, and at the same time, have the correct orientation (steric factor) for successful collision. The constant 'A' usually has the same value for small temperature changes. So, most of the time in kinetics study, it is assumed to be invariant with temperature.

3. The orders of reaction for the alkaline hydrolysis of two bromoalkanes are given below.

Reactants	Order with respect to bromoalkane	Order with respect to OH^-
(I) $CH_3CH_2CH_2Br$ and OH^-	1	1
(II) $(CH_3)_3CBr$ and OH^-	1	0

(a) In the context of reaction (I), write down the rate equation and explain what is meant by the terms *rate of reaction*, *rate constant*, and *overall order*.

Explanation:

The rate equation is rate $= k[CH_3CH_2CH_2Br][OH^-]$.

The rate equation is a mathematical equation that shows how the rate of reaction is dependent on the concentrations of both $CH_3CH_2CH_2Br$ and OH^-. In this case, the reaction is first order with respect to both $[CH_3CH_2CH_2Br]$ and $[OH^-]$, respectively.

The rate constant, k, is a proportionality constant in the rate equation of the reaction.

The overall order of reaction is the sum of the powers to which the concentrations of the reactants are raised to in the rate law.

(b) Explain the effect on the rate of each reaction if only the concentration of hydroxide ion is doubled.

Explanation:

If $[OH^-]$ is doubled, being a first-order reaction with respect to $[OH^-]$, the rate of reaction would be doubled.

(c) Briefly explain how you would study the rate of one of these reactions in the laboratory.

Explanation:

Using the continuous method, we can monitor the pH of the reaction mixture as time progresses. The procedures are:

(1) Mix 20 cm^3 of $[CH_3CH_2CH_2Br] = y$ mol dm^{-3} and 20 cm^3 of $[OH^-] = x$ mol dm^{-3}. Make sure that $y >> x$.
(2) Monitor the pH of the mixture from $t = 0$ at regular time intervals till the reaction is almost completed.
(3) Next, mix another 20 cm^3 of $[CH_3CH_2CH_2Br] = 2y$ mol dm^{-3} and 20 cm^3 of $[OH^-] = x$ mol dm^{-3}. Make sure that $y>>x$.
(4) Monitor the pH of the mixture from $t = 0$ at regular time intervals till the is reaction almost completed.
(5) Plot two graphs of pH versus time.

Q How do we deduce the order of the reaction with respect to $[CH_3CH_2CH_2Br]$ and $[OH^-]$?

A: Since $pH = -\log [H^+] = 14 + \log [OH^-]$, as $[OH^-]$ decreases, the pH of solution would also decrease.

To find the order with respect to $[OH^-]$:

(1) If the pH versus time plot is a decreasing linear straight line ⇒ reaction is a zero-order reaction.
(2) If it is a decreasing curve, then find out whether the half-life is constant. If it is, a first-order reaction.
(3) If not, at different concentration points on the curve, draw tangents and calculate the gradients. Each gradient is the rate at a particular concentration value. Then plot rate versus $[OH^-]^2$, which is equivalent to rate versus $(10^{-(14-pH)})^2$. If the plot is a straight line, it is a second-order reaction.

To find the order with respect to $[CH_3CH_2CH_2Br]$:

(1) If the two graphs for $[CH_3CH_2CH_2Br] = y$ mol dm^{-3} and $2y$ mol dm^{-3} are exactly the same line, then the reaction is zero order.
(2) If the two graphs for $[CH_3CH_2CH_2Br] = y$ mol dm^{-3} and $2y$ mol dm^{-3} are two different lines, then draw tangents at $t = 0$ for the two graphs and determine the initial rate values. Compare the two values and observe to what extent the rate has changed when $[CH_3CH_2CH_2Br]$ is doubled.

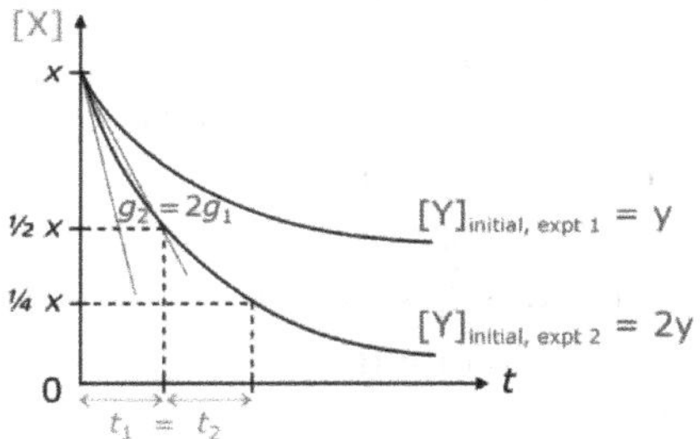

Q Why is $y >> x$?

A: Since we are monitoring the $[OH^-]$ with respect to time, to ensure that the change of $[OH^-]$ is affected minimally by the $[CH_3CH_2CH_2Br]$, we have to use $[CH_3CH_2CH_2Br]$ in large excess as compared to $[OH^-]$. Under such conditions, the rate law is approximated to a pseudo-*n*th order reaction with respect to $[OH^-]$:

$$\text{rate} = k[OH^-]^n\,[CH_3CH_2CH_2Br]^m.$$

If $[CH_3CH_2CH_2Br] >> [OH^-]$, then $[CH_3CH_2CH_2Br] \approx$ constant as the reaction progresses. Thus,

$$\text{rate} = k'[OH^-]^n, \text{ where } k' = k[CH_3CH_2CH_2Br]^m.$$

This is the method we normally would use in kinetics study when trying to monitor the concentration of a species. We make sure that the rate of reaction at each point corresponds to the concentration of the species that we are monitoring, and is not influenced by the concentration of the other species!

If $b >> a$, and the $[CH_3CH_2CH_2Br]$ is later doubled to $2b$, shouldn't the rate be unaffected since the $[CH_3CH_2CH_2Br]$ is already so high?

A: This is a common misconception that a lot of students have. As long as the order is not zero with respect to a reactant, increasing the concentration would certainly increase the rate. This is logical as a higher concentration means a greater frequency of collision in accordance to the collision theory of reaction.

(d) In one such experiment on reaction (II), it was found that the time taken for the concentration of $(CH_3)_3CBr$ to fall to half of its starting concentration is 45 min. Determine the rate constant of the reaction and how long it would take for the concentration of $(CH_3)_3CBr$ to fall to 1/8 of its starting concentration. Explain your answer.

Explanation:

According to the data, the rate equation is rate $= k[(CH_3)_3CBr]$. Since this is a first-order reaction, the $t_{½}$ is a constant:

$$t_{½} = \frac{\text{In}\,2}{k} = \frac{\text{In}\,2}{k} = 45 \Rightarrow k = 1.54 \times 10^{-2}\ \text{min}^{-1}$$

Concentration at nth half-life is $C = \left(\frac{1}{2}\right)^n . C_o$ where C_o is the original concentration.

So, for the concentration of $(CH_3)_3CBr$ to fall to 1/8 of its starting concentration, three half-lives have passed.

Therefore, the time needed is $= 3 \times 45 = 135$ min.

4. (a) The first step in a possible mechanism for the reaction $2NO(g) + O_2(g) \rightarrow 2NO_2(g)$ is as follows:

$$2NO \rightleftharpoons N_2O_2.$$

(i) Draw the dot-and-cross diagrams of NO and N_2O_2. What feature of the electronic structure of NO suggests that this is a likely first step in this reaction?

Explanation:

The N atom of NO does not have an octet configuration and it has a lone electron. The NO molecule is thus an electron-deficient radical. Hence, the formation of N_2O_2 is an exothermic process, which means that the first step is unlikely to be a slow step. If the other mechanistic step after the first step is a slow step, then each of the steps before the slow step would reach an equilibrium state.

Do you know?

— If you need to select the slow step from two given steps, then the one that has a more endothermic enthalpy change is likely to be the slow step. Usually, comparing a step that involves bond–breaking with a step that has only bond formation, the bond-breaking step is a slow step simply because bond-breaking needs more activation energy than a bond-forming reaction.

(ii) Explain why the enthalpy change for this step is -163 kJ mol^{-1}, given that the average bond energy for the N–N bond in compounds of nitrogen is $+163$ kJ mol^{-1}.

Explanation:

This particular step involves the formation of a N–N bond only. The amount of energy that is released for forming this bond must be equivalent to the amount of energy needed to break this bond. Hence, this accounts for why the enthalpy change for this step is −163 kJ mol^{-1}, given that the average bond energy for the N-N bond in compounds of nitrogen is +163 kJ mol^{-1}.

(iii) Explain why this step does not control the rate of the reaction. Assuming there is one further step in the reaction, write an equation for this.

Explanation:

A mechanistic step that is exothermic in nature does not control the rate of the reaction. The next step is likely to be: $N_2O_2 + O_2 \rightarrow 2NO_2$, which is the r.d.s. It is likely to be the slow step due to the involvement of bond-breaking during the reaction.

(b) (i) Sketch the Maxwell–Boltzman distribution of molecular energies at a temperature T_1 K. On the same axes, sketch the curve which shows the distribution of molecular energies at a higher temperature T_2 K.

Explanation:

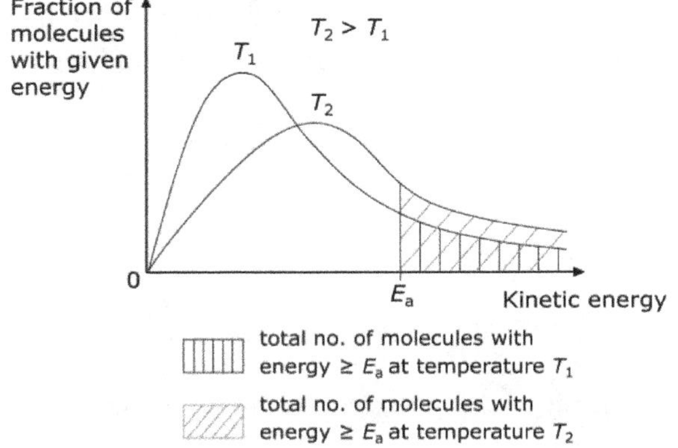

(ii) Use these graphs to explain how the rate of a gas phase reaction changes with increasing temperature.

Explanation:

When we increase the temperature of the system, the rate of reaction increases because at a higher temperature, (i) we now have more particles with the kinetic energy greater than the activation energy; this leads to higher frequency of effective collisions, and (ii) since the particles have more kinetic energy, they move faster, hence the higher frequency of collisions would lead to more effective collisions.

(c) For a gaseous reaction, state and explain what effects the addition of a catalyst would have on:

(i) the energy distribution of the gas molecules;

Explanation:

The addition of a catalyst has no effect on the energy distribution of the gas molecules.

Do you know?

— According to the law of thermodynamics, the energy of a system can only be changed through heat flow or work done.

(ii) the activation energy for the reaction;

Explanation:

The addition of a catalyst lowers the activation energy for the reaction as a catalyzed reaction has a lower E_a as compared to an uncatalyzed one.

Do you know?

— There are two types of catalysis, homogeneous and heterogeneous catalysis.
Homogeneous catalysis:
The reaction between I^-(aq) and $S_2O_8^{2-}$ (aq) is catalyzed by Fe^{3+}:

Step 1: $2I^- + 2Fe^{3+} \rightarrow I_2 + 2Fe^{2+}$
Step 2: $2Fe^{2+} + S_2O_8^{2-} \rightarrow 2SO_4^{2-} + 2Fe^{3+}$
Overall: $2I^-\,(aq) + S_2O_8^{2-}\,(aq) \rightarrow I_2(aq) + 2SO_4^{2-}\,(aq)$

The catalyzed reaction proceeds via a two-step route. Each step involves a reaction between oppositely charged ions which have lower E_a than the overall reaction, which involves the collision of two particles of the same charge. This lowers the E_a and enhances the reaction rate as it facilitates the ease of electron transfer in an oxidation–reduction reaction. ($E^{\theta}_{cell} > 0$ for each step of the reaction.)
(Note: If Fe^{2+} is added initially, the catalyzed reaction will begin with Step 2 followed by Step 1.)

(Continued)

(*Continued*)

Heterogeneous catalysis:
In heterogeneous catalysis, the reactants adsorb on the active sites of the catalyst surface. The catalyst lowers down the activation energy by:

- orientating reactant particles so that they achieve the correct collision geometry;
- locally increasing concentrations of the reactant particles; and
- weakening the intramolecular bonds of the reactant molecules.

(iii) the rate of reaction;

Explanation:

Since the catalyzed reaction has a lower activation energy, the rate of reaction would increase. This is reflected in a greater rate constant, k, for a catalyzed reaction.

Do you know?

— From Arrhenius rate constant, $k = \text{Aexp}\left(-\frac{E_a}{RT}\right)$, k increases when E_a decreases.

(iv) the overall order of reaction for both homogeneous and heterogeneous catalysis; and

Explanation:

For homogeneous catalysis, an example is the reaction between I^- (aq) and $S_2O_8^{2-}$ (aq) which is catalyzed by Fe^{3+}. The rate law for an uncatalyzed reaction is rate = $k[I^-][S_2O_8^{2-}]$. During catalysis by Fe^{3+}, the first step becomes $2I^- + 2Fe^{3+} \rightarrow I_2 + 2Fe^{2+}$ and if this is the r.d.s., then the rate law is rate = $k[I^-][Fe^{3+}]$. But if it is the second step that is the r.d.s., then the rate law may be rate = $k[I^-][S_2O_8^{2-}][Fe^{3+}]$. So, the overall order of reaction may change for homogeneous catalysis.

As for heterogeneous catalysis, the reactants adsorb on the active sites, and if we treat an active site as a kind of reactant, AS, then the fact that the active site is involved in the r.d.s. would mean we should include it in the rate law. Hence, the overall order of reaction may change for heterogeneous catalysis, i.e., rate = k[reactant][AS] instead of rate = k[reactant].

(v) the individual orders for both homogeneous and heterogeneous catalysis.

Explanation:

Based on the explanation in part (*iv*), depending on the reaction mechanism of a homogeneously catalyzed reaction and which one of the few elementary steps is the slow step in the mechanism, the individual order may change. As for heterogeneous catalysis, because the particles reside on the active site throughout the reaction, the individual order may also change.

(d) Thioethanamide reacts with sodium hydroxide as follows:

$$CH_3CSNH_2 + 2OH^- \rightarrow CH_3CO_2^- + HS^- + NH_3.$$

The reaction is first order with respect to both thioethanamide and hydroxide ions.

(i) Write the rate equation for this reaction.

Explanation:

The rate equation is rate = $k[CH_3CSNH_2][OH^-]$.

(ii) Given that the reaction occurs in two stages and the r.d.s. is

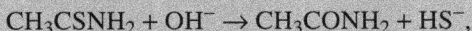

$$CH_3CSNH_2 + OH^- \rightarrow CH_3CONH_2 + HS^-,$$

write an equation for the second step in the reaction.

Explanation:

Step 1: $CH_3CSNH_2 + OH^- \rightarrow CH_3CONH_2 + HS^-$ Slow
Step 2: $CH_3CONH_2 + OH^- \rightarrow CH_3CO_2^- + NH_3$ Fast

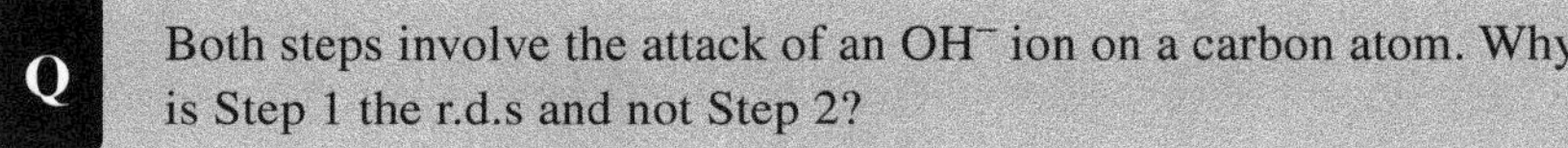

Q Both steps involve the attack of an OH^- ion on a carbon atom. Why is Step 1 the r.d.s and not Step 2?

A: In an organic reaction, the readiness of attack depends on the attractive force of the two oppositely charged centers. In CH_3CSNH_2, the C is bonded to a N and a S atom, while for CH_3CONH_2, the C is bonded to a N and an O atom. We would expect the carbon atom of CH_3CONH_2 to be more electron-deficient, hence attracting the OH^- ion more readily. Therefore, a lower activation energy is needed.

CHAPTER 6

CHEMICAL EQUILIBRIA

1. Ethanol, which is an important motor fuel nowadays, can be manufactured by the direct catalytic hydration of ethene with steam, using phosphoric acid as a catalyst. Assume that the reaction of ethene (C_2H_4) with steam to give ethanol (C_2H_5OH, which is gaseous at the temperature of the reaction) is at equilibrium.

   (a) Write the equation for the reaction (with state symbols).

Explanation:

$$C_2H_4(g) + H_2O(g) \rightleftharpoons C_2H_5OH(g).$$

(b) Write the expression for the *equilibrium constant*, K_c, for the reaction in terms of the concentrations of reactants and products.

Explanation:

$$C_2H_4(g) + H_2O(g) \rightleftharpoons C_2H_5OH(g), \qquad K_c = \frac{[C_2H_5OH]}{[C_2H_4][H_2O]}.$$

Do you know?

— When a system is at dynamic equilibrium, the concentration of each of the species in the chemical equation remains the same with the progress of time. But there is still some reactions taking place at the microscopic level, just that the rate of the forward reaction equals to the rate of the backward reaction. Thus, we can define a physical quantity called equilibrium constant to describe this dynamic system.

— An equilibrium constant must go together with an associating chemical equation. Chemical equations consisting of the same types of reactants and products but with different stoichiometric ratios would have different equilibrium constant. This is because the equilibrium constant is defined as the ratio of the product of the concentration of the products over the product of the concentration of the reactants. Each concentration term is raised to a power value in accordance to what the stoichiometric coefficient in the chemical equation is being depicted.

— Take note that an equilibrium constant can only be changed by temperature and not other variables such as pressure, concentration, or catalyst. This is because the equilibrium constant can also be expressed in terms of the ratio of the rate constants:

$$K_c = \frac{k_f}{k_b} = \frac{Ae^{-E_{a,f}/RT}}{Ae^{-E_{a,b}/RT}}.$$

(c) Use the information below to calculate the equilibrium concentration of ethanol vapor under these conditions:

Temperature = 570 K
Pressure = 60×10^5 Pa
K_c = $24 \text{ dm}^3\text{mol}^{-1}$ at 570 K
$[H_2O(g)]$ = $0.050 \text{ mol dm}^{-3}$
$[C_2H_4(g)]$ = 0.45 mol dm^{-3}

Explanation:

Given that $K_c = \frac{[C_2H_5OH]}{[C_2H_4][H_2O]} = \frac{[C_2H_5OH]}{(0.05)(0.45)} = 24$,
therefore, $[C_2H_5OH] = 0.54$ mol dm^{-3}.
(Note that the K_c is defined for temperature of 570 K.)

> (d) The enthalpy change for the forward reaction is $\Delta H = -46$ kJ mol^{-1}. Why is a temperature of 570 K used for the reaction, rather than
>
> (i) a higher temperature?

Explanation:

For an exothermic reaction, if the temperature is raised, according to Le Chatelier's Principle, the system would respond by taking in the excess heat energy. Hence, the position of equilibrium will shift left, i.e., toward the reactant side. Therefore, the yield of the product will decrease. For this particular reaction, a temperature that is higher than 570 K is not the optimal working temperature to get a good yield of product.

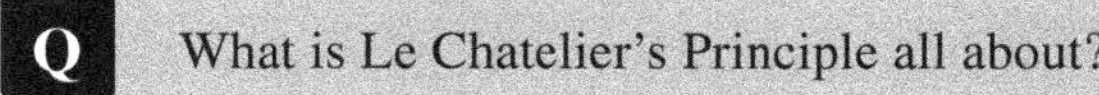

Q
What is Le Chatelier's Principle all about?

A: Le Chatelier's Principle is a very useful rule to predict how the position of an equilibrium would respond when an external strain is imposed onto the system. It simply states that "when a change is imposed on a system at equilibrium, the system would respond in such a way as to remove the change."

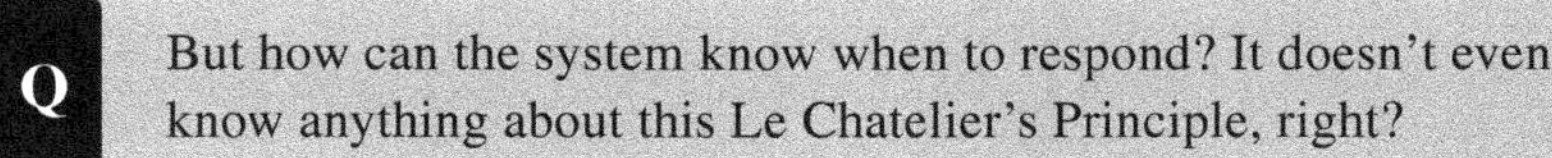

Q
But how can the system know when to respond? It doesn't even know anything about this Le Chatelier's Principle, right?

A: We need to use kinetics to understand how the system responds. Take for instance, the increase in temperature for an exothermic reaction. When the temperature is increased, from what we know in kinetics (Chapter 5), the rate of reaction would increase. Therefore, the rates for both the forward and backward reactions are going to increase. But the one that has a higher activation energy (since this is an exothermic reaction, it would be the backward reaction) is going to be increased by a greater percentage than the forward one. Hence, as time passes by, the position of equilibrium would shift to the left, i.e., to the reactant side.

Q So, does it mean that when we decrease the temperature for an endothermic reaction, according to Le Chatelier's Principle, the system has to respond in such a way to produce more heat? Hence, the position of equilibrium would shift to the left since the backward reaction is exothermic?

A: Absolutely spot on!

Is there a way to prove it?

A: Look at the following calculation, if we assume that the pre-exponential factor of the Arrhenius rate constant is 1:

E_a/kJ mol^{-1}	k = Aexp(−Ea/RT) at T = 300 K	k = Aexp(−Ea/RT) at T = 400 K	Percentage change in rate constant
500 (backward)	0.818	0.860	5.13
200 (forward)	0.923	0.942	2.06

Well, the rate constant for the higher E_a value has a larger percentage increase in value, although the rate constant for a lower E_a value is in fact greater than that for the higher E_a.

(ii) a lower temperature?

Explanation:

For an exothermic reaction, if the temperature is lowered, according to Le Chatelier's Principle, the system would respond by producing more heat energy. Hence, the position of equilibrium will shift to the right, i.e., toward the product side. Therefore, the yield of the product will increase.

For this particular reaction, a temperature that is lower than 570 K is not the optimal working temperature to get a good yield of product as the rate of reaction may not be fast enough.

Do you know?

— A change in temperature (1) changes the rate of reaction; (2) changes the position of equilibrium; and (3) changes the K_c.
For an exothermic reaction:
- Increase in temperature favors the backward reaction (endo); decreases K_c.
- Decrease in temperature favors the forward reaction (exo); increases K_c.

For an endothermic reaction:
- Increase in temperature favors the forward reaction (endo); increases K_c.
- Decrease in temperature favors the backward reaction (exo); decreases K_c.

(e) Why is a pressure of 60×10^5 Pa used for the reaction, rather than
(i) a higher pressure?

Explanation:

$$C_2H_4(g) + H_2O(g) \rightleftharpoons C_2H_5OH(g)$$

For a reaction that has more gas particles on the left than on the right, when the pressure is increased, according to Le Chatelier's Principle, the system would respond by producing less particles. Hence, the position of equilibrium will shift to the right, i.e., toward the product side. Therefore, the yield of the product will increase. For this particular reaction, a pressure that is higher than 60×10^5 Pa is not the optimal working pressure as more money needs to be invested in building a stronger reactor.

Q From the kinetics perspective, how does an increase in pressure shifts the position of equilibrium to the right?

A: When the total pressure increases due to a decrease in the volume of the container, the volume of the system shrinks. The gaseous particles now occupy a smaller volume. As a result, the effective collisional frequencies of both the forward and backward reactions increase. These lead to increases in the rates of both the forward and backward reactions. BUT, the rate of the reaction that involves a greater number of particles colliding is increased by a greater extent. So, in this case, since the forward reaction involves the collision of two particles, whereas the backward reaction involves only one, the percentage increase of the forward rate is higher than that of the backward rate. As time passes, the forward rate and backward rate become equal again, but at a higher value than the previous equilibrium state. Take for instance, the reaction of $PCl_3(g) + Cl_2(g) \rightleftharpoons PCl_5(g)$ below:

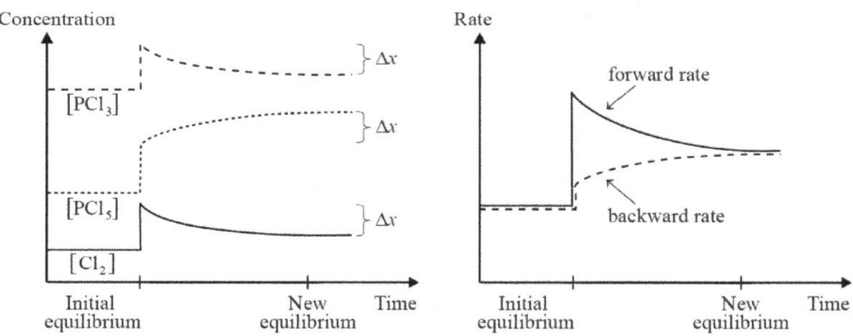

Note: Δx = change in [substance]

Did you notice that the change that is being imposed onto the system has not been totally removed even after the system reach a new equilibrium state? This is an important characteristic of an equilibrium system that you need to know.

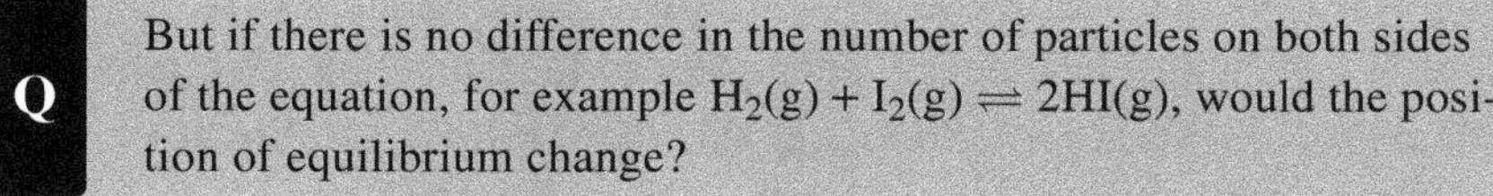

Q But if there is no difference in the number of particles on both sides of the equation, for example $H_2(g) + I_2(g) \rightleftharpoons 2HI(g)$, would the position of equilibrium change?

A: No! The position of equilibrium would not change, BUT the rate at the higher pressure would be greater, which is logical because the volume is smaller now, hence the particles have greater effective collision frequency.

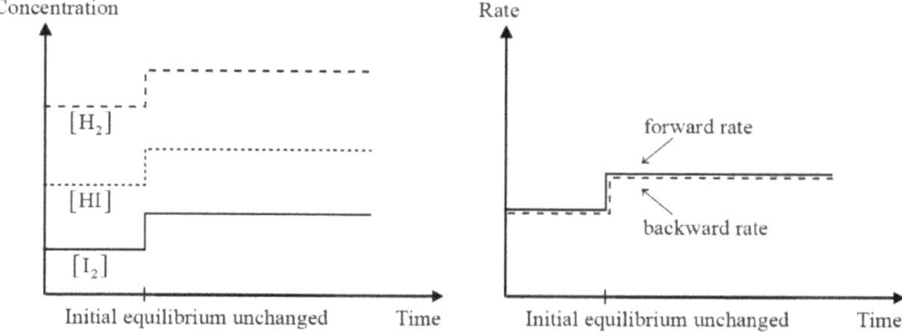

Q But the pressure of a system can also be increased by adding more reactant or product particles while keeping the volume the same, so how would the system respond to it?

A: If the volume does not change and we put in more reactant particles, this is equivalent to increasing the concentration of this particular reactant particles. Then according to Le Chatelier's Principle, the system would response in such a way to remove this excess reactant. The position of equilibrium will shift to the right, i.e., toward, the product side. Although the amount of products increases, the equilibrium constant still does not change. Take for instance, the reaction of $PCl_3(g) + Cl_2(g) \rightleftharpoons PCl_5(g)$ below:

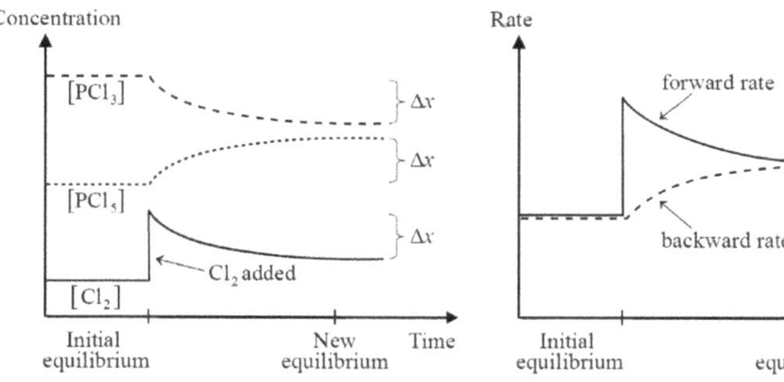

Note: Δx = change in [substance]

The following graphs show how the system would respond if some PCl_5 is being removed instead:

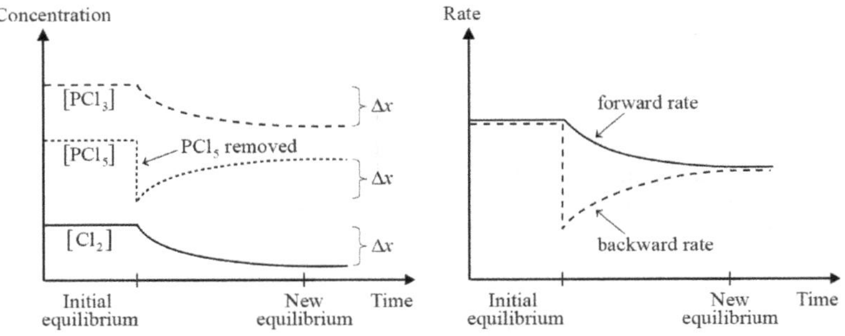

Note: Δx = change in [substance]

According to Le Chatelier's Principle, the system would respond by producing more PCl_5, hence the forward reaction would be favored and the position of equilibrium would shift to the right. From the kinetics perspective, the position of equilibrium would shift to the right is because when PCl_5 is being removed from the system at equilibrium, the backward rate is immediately decreased. On the other hand, the forward rate would still be maintained initially and then decreased. So as time passes by, the system would reach a new equilibrium state with a lesser amount of each particle as compared to the original equilibrium.

Q If we add in inert gas particles while maintaining a constant volume, how would the system respond to such change?

A: Since the pressure increases because of the addition of inert gas and the volume remains the same, the system would not shift its position of equilibrium as the inert gas particles do not affect the effective collsional frequency of the reacting particles.

Q But what would happen if the inert gas is added under a constant pressure condition?

A: According to pV = nRT, if n increases while keeping p and T constant, then V would have to increase. Therefore, the concentration of all the species in the system would immediately decrease. There are two different scenarios as shown below:

(1) The number of gas particles are not the same on both sides of the equation.

$$PCl_3(g) + Cl_2(g) \rightleftharpoons PCl_5(g)$$

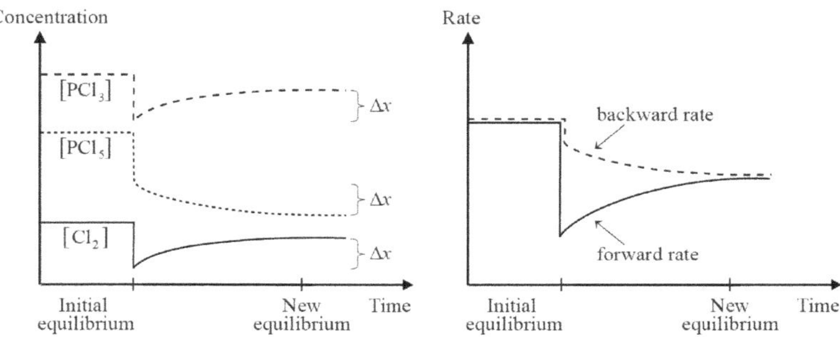

According to Le Chatelier's Principle, the system would attempt to counteract the change by favoring the reaction that increases the overall number of gaseous molecules, i.e., the backward reaction is favored. The equilibrium position will shift to the left. More $PCl_5(g)$ will dissociate to form $PCl_3(g)$ and $Cl_2(g)$ until a new equilibrium is established. The concentrations of all species (both reactant sand products) decrease. In terms of the number of moles, the new equilibrium mixture has a higher percentage of $PCl_3(g)$ and $Cl_2(g)$ but a lower percentage of $PCl_5(g)$ as compared to the old equilibrium position. The equilibrium constant remains unchanged.

(2) The number of gas particles are the same on both sides of the equation.

$$H_2(g) + I_2(g) \rightleftharpoons 2HI(g)$$

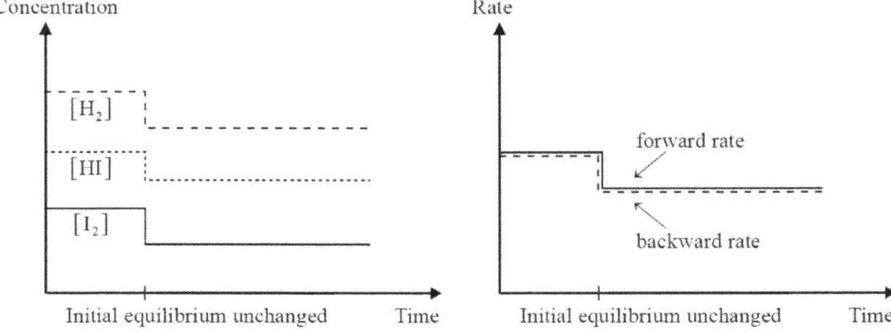

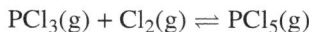

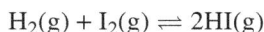

For this scenario, since both the forward and backward reactions produce the same number of gaseous molecules, none of these reactions is favored over the other. Thus, the equilibrium position will not shift as the system remains at equilibrium. In terms of the number of moles, the equilibrium composition is unchanged and so is the equilibrium constant. How would you use the kinetics perspective to rationalize it?

(ii) a lower pressure?

Explanation:

$$C_2H_4(g) + H_2O(g) \rightleftharpoons C_2H_5OH(g).$$

For a reaction that has more gas particles on the left side of the equation than on the right, when the pressure is decreased, according to Le Chatelier's Principle, the system would respond by producing more particles. Hence, the position of equilibrium will shift to the left, i.e., toward the reactant side. Therefore, the yield of the product will decrease. For this particular reaction, a pressure that is lower than 60×10^5 Pa is not the optimal working pressure as it would affect the yield.

Do you know?

— Adding reactant at constant volume, example: $PCl_3(g) + Cl_2(g) \rightleftharpoons PCl_5(g)$:
⇒ Change in rate of reaction;
⇒ Change in position of equilibrium;
⇒ No change in K_c.

— Removing reactant at constant volume, example: $PCl_3(g) + Cl_2(g) \rightleftharpoons PCl_5(g)$:
⇒ Change in rate of reaction;
⇒ Change in position of equilibrium;
⇒ No change in K_c.

— Compression of a system in which the number of gas molecules on each side of the chemical equation is different, example: $PCl_3(g) + Cl_2(g) \rightleftharpoons PCl_5(g)$:
⇒ Change in rate of reaction;

(Continued)

(*Continued*)

⇒ Change in position of equilibrium;
⇒ No change in K_c.

— Compression of a system in which the number of gas molecules on each side of the chemical equation is the same, example: $H_2(g) + I_2(g) \rightleftharpoons 2HI(g)$:
⇒ Change in rate of reaction;
⇒ No change in position of equilibrium;
⇒ No change in K_c.

— Adding inert gas at constant pressure to a system in which the number of gas molecules on each side of the chemical equation is different, example: $PCl_3(g) + Cl_2(g) \rightleftharpoons PCl_5(g)$:
⇒ Change in rate of reaction;
⇒ Change in position of equilibrium;
⇒ No change in K_c.

— Adding inert gas at constant pressure to a system in which the number of gas molecules on each side of the chemical equation is the same, example: $H_2(g) + I_2(g) \rightleftharpoons 2HI(g)$:
⇒ Change in rate of reaction;
⇒ No change in position of equilibrium;
⇒ No change in K_c.

— Adding inert gas at constant volume:
⇒ No change in rate of reaction;
⇒ No change in position of equilibrium;
⇒ No change in K_c.

2. The Haber process is an important industrial process for the synthesis of ammonia, a key precursor for making industrial fertilizers. Iron is a common catalyst used in this process. The process does not go to completion on its own and would reach the following dynamic equilibrium state:

$$N_2(g) + 3H_2(g) \rightleftharpoons 2NH_3(g).$$

(a) Define the terms *partial pressure* of a gas in a mixture of gases and *dynamic equilibrium*.

Explanation:

The *partial pressure* of a gas is the pressure exerted by this particular gaseous particles on the wall of the container. It is actually a fraction of the total pressure, i.e., $p_1 = x_1 . p_T$.

The term *dynamic equilibrium* refers to the state of a system when the rate of forward reaction is equal to the rate of backward reaction. As a result, no macroscopic changes are observed; both [reactants] and [products] remain constant. But at the microscopic level, both forward and backward reactions still proceed.

Do you know?

— An equilibrium can only be established for a closed system, one in which matter is not allowed to leave or enter the system.

— You can start off with different initial concentrations of the species; the system would be able to adjust itself to reach a dynamic equilibrium state with the same equilibrium constant; for example, consider the reaction: $A + B \rightleftharpoons C + D$:

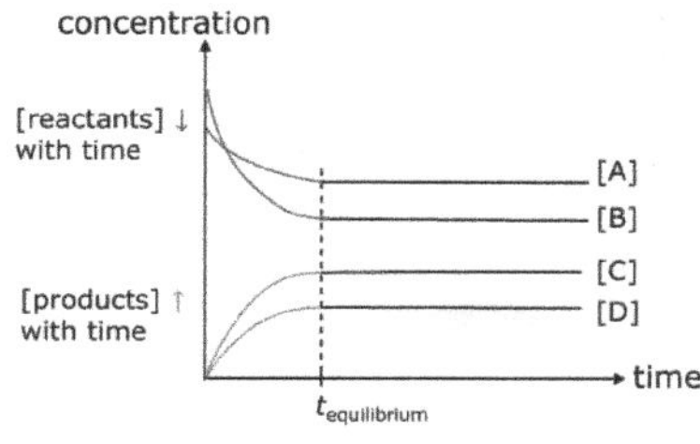

(b) In an equilibrium mixture consisting of nitrogen, hydrogen, and ammonia at 800 K, the partial pressure of the three gases are 20.4, 57.2, and 11.7 atm, respectively.

(i) Calculate the total pressure in the system.

Explanation:

According to Dalton's Law of Partial Pressure, the total pressure of a system is the sum of its partial pressures:

$$P_T = p_{N_2} + p_{H_2} + p_{NH_3} = 20.4 + 57.2 + 11.7 = 89.3 \text{ atm}$$

(ii) What is the mass of ammonia present under these conditions in a vessel of volume 200 m^3? [Take $R = 8.20 \times 10^{-5}$ m^3 atm mol^{-1} K^{-1}]

Explanation:

Assume that the ammonia is an ideal gas,

$$pV = nRT = \frac{(\text{mass})RT}{\text{molar mass}} \Rightarrow \text{Mass} = \frac{\text{Molar mass} \times p \times V}{RT}$$

$$= \frac{17 \times 11.7 \times 200}{800 \times 8.20 \times 10^{-5}} = 606402.4\text{g}$$

$$= 6.06 \times 10^5 \text{ g.}$$

(iii) Write an expression for K_p for the formation of ammonia.

Explanation:

$$N_2(g) + 3H_2(g) \rightleftharpoons 2NH_3(g), \quad K_p = \frac{(p_{NH_3})^2}{(p_{N_2}) \cdot (p_{H_2})^3}.$$

Do you know?

— Like K_c, K_p is ONLY temperature dependent.
— The relationship between K_p and K_c is:
Since $pV = nRT \Rightarrow p = [\text{species}]RT$,
therefore, $K_p = \frac{(p_{NH_3})^2}{(p_{N_2}).(p_{H_2})^3} = \frac{([NH_3]RT)^2}{([N_2]RT)([H_2]RT)^3} = K_c \frac{1}{(RT)^2}$.

(iv) Calculate the value of K_p under these conditions.

Explanation:

$$K_p = \frac{(p_{NH_3})^2}{(p_{N_2}) \cdot (p_{H_2})^3} = \frac{(11.7)^2}{(20.4)(57.2)^3} = 2.05 \times 10^{-3}\,\text{atm}^{-2}.$$

(v) What would the effect on the equilibrium yield of ammonia be under the conditions in part *(b)(ii)* if a better catalyst were used?

Explanation:

Since a catalyst does not change the position of equilibrium but instead speeds up the rate to achieve equilibrium, the equilibrium yield of the ammonia would not be affected.

Do you know?

— Adding a catalyst would result in (1) change in rate of reaction; (2) no change in position of equilibrium; and (3) no change in K_c. This is because a catalyst lowers the activation energy of both the forward and backward reactions by the same amount, as shown below:

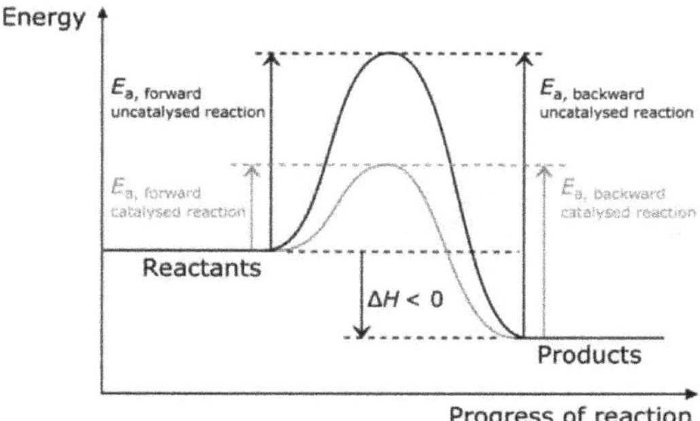

(vi) Predict and explain the effect of an increase in temperature on the value of K_p.

Explanation:

The formation of NH_3 is an exothermic process. According to Le Chatelier's Principle, when temperature is increased, the system would respond to remove the excess heat. Thus, the backward reaction would be favored and the position of equilibrium would shift left, forming less products. Hence, the value of the K_p would decrease.

Q How do you know that the reaction is exothermic in nature?

A: Well, you can use bond energy data to calculate:

$$\Delta H = [BE(N{\equiv}N) + 3BE(H{-}H)] - 6BE(N{-}H)$$
$$= 994 + 3(436) - 6(390) = -38 \text{ kJ mol}^{-1}.$$

Do you know?

— It would also mean that the value of the K_c would also decrease with increasing temperature. Vice versa, the K_p or K_c value would increase with decreasing temperature.

— The formation of ammonia is an exothermic process and a high temperature is used because of the need to break the strong N≡N triple bond as, the activation energy is high.

(vii) Predict and explain the effect of an increase in temperature on the rate of the forward reaction.

Explanation:

An increase in the temperature would increase the rate of the forward reaction. This is because at a higher temperature, (i) we now have more particles with the kinetic energy greater than the activation energy; this leads to higher frequency of effective collisions, and (ii) since the particles have more kinetic energy, they move faster, hence the higher frequency of collisions would lead to more effective collisions. The following Maxwell–Boltzmann distribution plot reinforces the explanation:

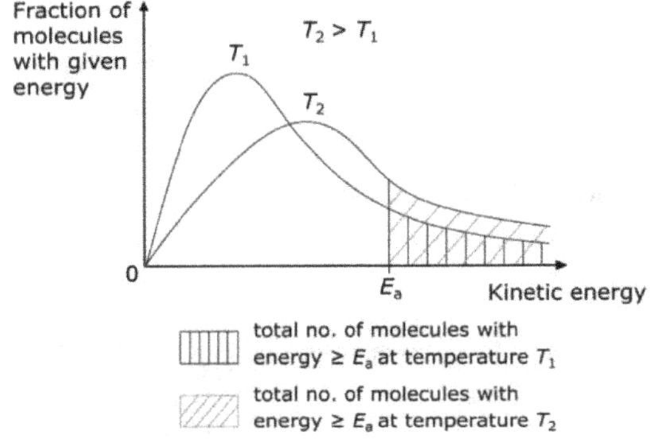

(viii) Give two reasons why a new catalyst might be preferred to the existing one even though it costs more.

Explanation:

If the new catalyst helps the system to achieve equilibrium faster, then by forming the products faster, we can save time and use this extra time to get more products. In addition, if the new catalyst lower down the activation energy much more than the old one, then a lower operating temperature can be used. Hence, we can get more products at a lower temperature for the new catalyst as the reaction has an exothermic enthalpy change.

(c) The gases are passed through a conversion chamber containing granulated iron as a catalyst. Describe and explain the effect of the iron on:

(i) the rate of the production of ammonia, and

Explanation:

The iron catalyst only increases the rate to achieve equilibrium. Hence, it allows us to get the product faster.

(ii) the amount of ammonia in the equilibrium mixture.

Explanation:

The iron catalyst does not shift the position of the equilibrium. Hence, a catalyzed reaction would give us the same amount of product as an uncatalyzed one.

(d) The equilibrium mixture formed is then passed into a refrigeration plant. Explain why this is done and what happens after this stage.

Explanation:

By passing the equilibrium mixture into the refrigeration plant, the ammonia is being liquefied and drained off. This allows the unreacted H_2 and N_2 gases to be redirected back to the reaction chamber to be converted to ammonia. Hence, it does not cause any wastage to the unreacted reactants.

3. At high temperatures, phosphorus pentachloride is a gas that dissociates as follows:

$$PCl_5(g) \rightleftharpoons PCl_3(g) + Cl_2(g).$$

(a) Write an expression for the equilibrium constant, K_p, for this equilibrium.

Explanation:

$$PCl_5(g) \rightleftharpoons PCl_3(g) + Cl_2(g), \quad K_p = \frac{(p_{PCl_3})(p_{Cl_2})}{(p_{PCl_5})}.$$

Q Is the above reaction endothermic or exothermic?

A: The enthalpy change of the reaction = 2BE(P–Cl) – BE(Cl-Cl) = 2(79) – 244 = –86 kJ mol^{-1}. It is an exothermic process. So, how would you expect the position of equilibrium to shift when you decrease temperature? Try it out!

Q Since the above reaction involves bond breaking, shouldn't it be endothermic?

A: The forward reaction involves bond breaking only, but there is also bond formation on the product side. Using bond energy calculation indicates that it is exothermic in nature because the bond that is formed in the product is stronger than the bonds that are being broken in the reactant side. Thus, bond breaking process for the forward reaction can only be endothermic if there is no bond formation in the product side. Similarly, bond forming process for the forward reaction can only be exothermic if there is no bond breaking in the reactant side. The overall ethalpy change of a reaction is dependent on the sum of bond breaking versus bond forming.

(b) At a given temperature, the degree of dissociation of an original sample of $PCl_5(g)$ is 0.52 and the system reaches equilibrium. If the total equilibrium pressure is 2 atm, calculate the values of the equilibrium partial pressures of PCl_5 and PCl_3.

Explanation:

Let the initial amount of $PCl_5(g)$ be a moles.
Using the I.C.E. table:

	$PCl_5(g)$	$\rightleftharpoons$	$PCl_3(g)$	+	$Cl_2(g)$
Initial amount/mol	a		0		0
Change/mol	$-0.52a$		$+0.52a$		$+0.52a$
At equil./mol	$0.48a$		$+0.52a$		$+0.52a$

Total amount of gas particles at equilibrium = $0.48a + 0.52a + 0.52a = 1.52a$
Partial pressure of $PCl_5(g) = (0.48a/1.52a) \times 2 = 0.632$ atm
Partial pressure of $PCl_3(g)$ = partial pressure of $Cl_2(g) = (0.52a/1.52a) \times 2 = 0.684$ atm.

Q What is the degree of dissociation?

A: The degree of dissociation is the ratio of the amount dissociated over the original amount. It is a measurement of the extent of dissociation. The higher the degree of dissociation, the more the position of equilibrium is shifted to the right of the equation.

(c) Hence, calculate the value of K_p and give its units.

Explanation:

$$K_p = \frac{(P_{PCI_3})(P_{CI_2})}{(P_{PCI_5})} = \frac{0.684 \times 0.684}{0.632} = 0.741 \text{ atm.}$$

4. This question concerns the reaction

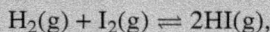

$$H_2(g) + I_2(g) \rightleftharpoons 2HI(g),$$

which is slow even at high temperature.

(a) (i) Using your Data Booklet, calculate ΔH for the reaction between hydrogen and iodine.

Explanation:

$$\begin{aligned} \Delta H &= \text{BE(H–H)} + \text{BE(I–I)} - 2\text{BE(H–I)} \\ &= 436 + 151 - 2(297) = -7 \text{ kJ mol}^{-1}. \end{aligned}$$

So, the reaction is slow because there is little difference between the energy levels of the reactant and product? Hence, there isn't much difference between the energetics stability of the reactants and product?

A: Yes, you are right. The H–I bond formed is weak; hence, the formation process is energetically not very favourable. The small enthalpy change also accounts for why the reaction is reversible as the difference in the forward and backward activation energies is small. So, take note that a reversible reaction is one whereby the enthalpy change is negligible.

(ii) Sketch an energy level diagram for this reaction.

Explanation:

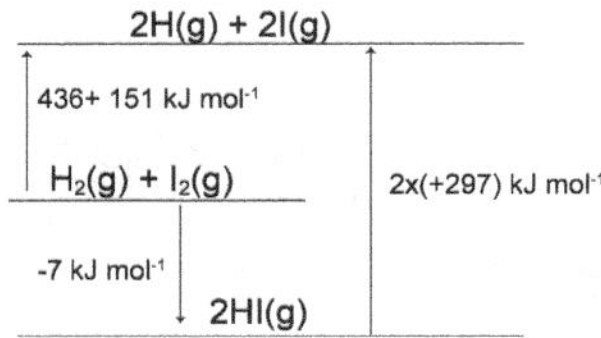

(b) Sketch the energy profile diagram and indicate on your sketch,

(i) ΔH for the reaction;

(ii) the activation energy for the forward reaction ($E_{a(f)}$); and

(iii) the activation energy for the reverse reaction ($E_{a(b)}$).

Explanation:

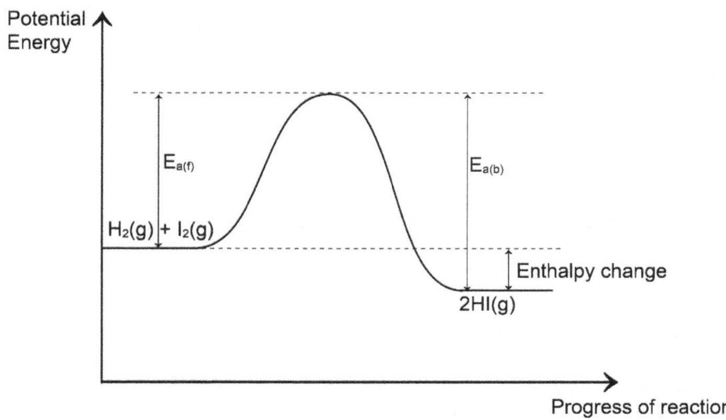

(c) For an analogous reaction involving the formation of hydrogen chloride

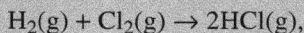

$$H_2(g) + Cl_2(g) \rightarrow 2HCl(g),$$

explain how you would expect the activation energy of the forward reaction to be compared with that shown for the formation of HI.

Explanation:

The activation energy of the forward reaction $H_2(g) + Cl_2(g) \rightarrow 2HCl(g)$ should be higher than that for the formation of HI(g) because more energy is needed to break the stronger Cl–Cl bond (BE = 244 kJ mol^{-1}).

> **Do you know?**
>
> — The Cl–Cl bond is stronger than the I–I bond because the orbital that the Cl atom uses to form the bond is smaller, hence it is less diffuse. Therefore, the overlap of the Cl–Cl bond is more effective than that of the I–I bond.

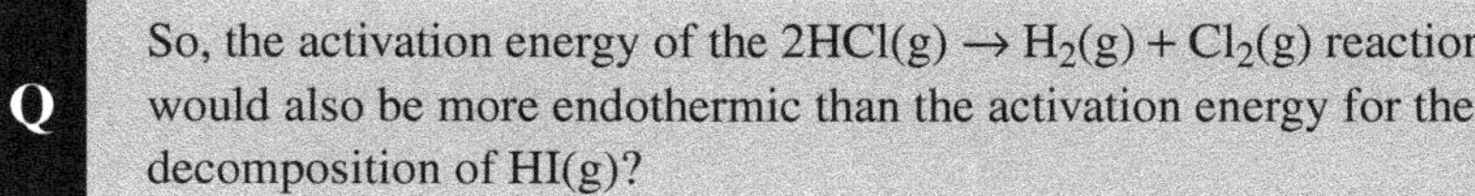

Q So, the activation energy of the $2HCl(g) \rightarrow H_2(g) + Cl_2(g)$ reaction would also be more endothermic than the activation energy for the decomposition of HI(g)?

A: Yes, indeed. This is because more energy is needed to break the stronger H-Cl bond due to greater effective overlap of the orbitals in the H–Cl bond as compared to that of the H–I bond.

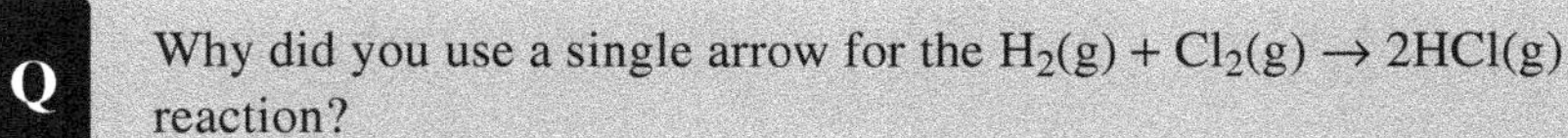

Q Why did you use a single arrow for the $H_2(g) + Cl_2(g) \rightarrow 2HCl(g)$ reaction?

A: This is because the reaction goes to completion. Hence, for a completed reaction, we use a single arrow.

> (d) The reaction for the formation of hydrogen iodide does not go to completion but reaches an equilibrium state.
>
> (i) Write an expression for the equilibrium constant, K_c, for this reaction.

Explanation:

$$H_2(g) + I_2(g) \rightleftharpoons 2HI(g), \quad K_c = \frac{[HI]^2}{[H_2][I_2]}.$$

(ii) A mixture of 2.9 mol of H_2 and 2.9 mol of I_2 is prepared and allowed to reach equilibrium in a closed vessel of 250 cm^3 capacity at 700°C. The resulting equilibrium mixture is found to contain 4.5 mol of HI. Calculate the value of K_c at this temperature.

Explanation:

Using the I.C.E. table:

	$H_2(g)$	+	$I_2(g)$	$\rightleftharpoons$	$2HI(g)$
Initial conc./mol dm^{-3}	$\frac{2.9}{0.250}$		$\frac{2.9}{0.250}$		0
Change/mol dm^{-3}	$-\frac{4.5}{0.250\times 2}$		$-\frac{4.5}{0.250\times 2}$		$+\frac{4.5}{0.250}$
At equil./mol dm^{-3}	2.6		2.6		18

$$K_c = \frac{[HI]^2}{[H_2][I_2]} = \frac{(18)^2}{2.6\times 2.6} = 2.66$$

(iii) Explain why the formation of hydrogen chloride goes to completion as compared to the formation of hydrogen iodide.

Explanation:

The HCl formed is more stable than HI due to the stronger H–Cl bond, thus the reaction for the formation of HCl goes to completion. In addition, because of the stronger H–Cl bond, the activation energy for the decomposition of HCl is more endothermic than that for the HI. Therefore, the decomposition of HCl is less likely to occur, which means the formation of HCI is less likely to be reversible.

Do you know?

— A reaction is likely to be reversible if the ΔH of reaction is approximately zero, meaning the activation energies for both the forward and backward reactions are about the same.

— In the case of HI, the ΔH of reaction for $H_2(g) + I_2(g) \rightleftharpoons 2HI(g)$ is -7 kJ mol^{-1}, which is not equal to zero. But because the activation energy of the backward reaction is not much different from that of the forward reaction, this would mean that at the reacting temperature in which HI can be formed, the decomposition of HI is also likely to happen. Hence, this accounts for the reversibility of the reaction.

(e) In an experiment to establish the equilibrium concentration in *(d)(ii)*, the reaction was allowed to reach equilibrium at 723 K and then quenched by addition to a known, large volume of water. The concentration of iodine in this solution was then determined by titration with standard sodium thiosulfate solution.

(i) Explain the purpose of quenching.

Explanation:

The purpose of quenching is to slow down the reaction. This is made possible here by adding water, i.e., increasing the volume of the solution. Hence, the reacting particles are farther apart and less likely to collide with one another.

(iii) Write an equation for the reaction between sodium thiosulfate and iodine.

Explanation:

$$I_2(aq) + 2S_2O_3^{2-} \rightarrow 2I^-(aq) + S_4O_6^{2-}(aq)$$

(iv) What indicator would you use? Give the color change when the endpoint is reached.

Explanation:

Starch solution is added to the solution when it turns pale yellow. A blue-black coloration is observed. The endpoint is reached when the blue-black decolorizes.

Do you know?

— The blue-black coloration is due to the formation of the starch–iodine complex. The starch solution cannot be added too early when the concentration of the iodine is still high as too much of the complex would form. The insolubility of the starch–iodine complex may prevent some of the iodine from reacting with the thiosulfate as they are embedded within the starch matrix.

(v) In this titration and in titrations involving potassium manganate (VII), a color change occurs during the reaction. Why is an indicator usually added in iodine/thiosulfate titrations but not in titrations that involve potassium manganate (VII)?

Explanation:

When the purple potassium manganate (VII) is used during titration, the purple MnO_4^- ion is converted to the near colorless Mn^{2+}. Hence, the endpoint can be easily distinguished when one drop of excess potassium manganate(VII) causes the solution to turn pink.

But for iodine thiosulfate titrations, toward the endpoint, the solution turns pale yellow. Hence, it is not easy to detect the color transition from pale yellow to colorless. Therefore, by creating a blue-black colouration, we can prominently detect the color transition from blue-black to colorless.

(f) The rate expression for the forward reaction between hydrogen and iodine is

$$\text{Rate} = k[H_2][I_2].$$

(i) What is the order of the reaction with respect to iodine?

Explanation:

The reaction is first order with respect to iodine.

(ii) When 0.20 mol of each of H_2 and I_2 were mixed at 600°C in a vessel of 500 cm^3 capacity, the initial rate of formation of HI was found to be 2.3×10^{-5} mol dm^{-3} s^{-1}. Calculate the value for k at 600°C, stating the units.

Explanation:

$$H_2(g) + I_2(g) \rightleftharpoons 2HI(g).$$

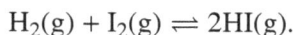

Given the rate of formation of HI = 2.3×10^{-5}mol dm^{-3} s^{-1},
the rate of depletion of I_2 = ½ × $2.3 \times 10^{-5} = 1.15 \times 10^{-5}$mol dm^{-3} s^{-1}.

$$\text{Rate} = k[H_2][I_2] = k\left(\frac{0.2}{0.5}\right)\left(\frac{0.2}{0.5}\right) = 1.15 \times 10^{-5}\text{mol dm}^{-3}\text{ s}^{-1}$$
$$\Rightarrow k = 7.19\times10^{-5}\text{mol}^{-1}\text{ dm}^3\text{ s}^{-1}.$$

Why did you need to half the rate of formation of HI when you calculate the "rate" for the rate equation? Can't we just use the rate of formation of HI directly?

A: Given a reaction: $aA + bB \rightarrow cC + dD$,
rate of formation of D = $+\frac{d[D]}{dt}$; rate of formation of C = $+\frac{d[C]}{dt}$;
rate of depletion of B = $-\frac{d[B]}{dt}$; rate of depletion of A = $-\frac{d[A]}{dt}$.
But rate = $k[A]^n[B]^m$, the rate $= -\frac{1}{a}\frac{d[A]}{dt} = \frac{1}{b}\frac{d[B]}{dt} = +\frac{1}{c}\frac{d[C]}{dt} = +\frac{1}{d}\frac{d[D]}{dt}$.
Hence, rate $= -\frac{1}{1}\frac{d[I_2]}{dt} = +\frac{1}{2}\frac{d[HI]}{dt}$.

Q Do we need to memorize the units for the various rate constant, k, of all the different orders of reaction?

A: You don't have to, just remember that the unit for rate is mol dm^{-3} $time^{-1}$, then derive as follows:

$$\text{Rate} = k[A]^m[B]^n$$
$$\Rightarrow \text{mol dm}^{-3}\ \text{time}^{-1} = k(\text{mol dm}^{-3})^n\ (\text{mol dm}^{-3})^m$$
$$\Rightarrow k = \text{mol}^{1-n-m}\ \text{dm}^{-3\ (1-n-m)}\ \text{time}^{-1}.$$

CHAPTER 7

IONIC EQUILIBRIA

1. It is now generally accepted that the acidic gas, sulfur dioxide, is one of the principal causes of acid rain. Atmospheric sulfur dioxide dissolves in water droplets in the air to form an acidic solution of low pH, which damages many vegetation and contaminates water bodies when it falls as rain.

   (a) Define

      (i) acid, and

Explanation:

According to Brϕnsted–Lowry definition, an acid is a proton donor.

Do you know?

— According to Arrhenius, an acid is a substance that dissociates in water to form hydrogen ions, H^+. While a base is a substance that dissociates in water to form hydroxide ions, OH^-.

— According to Brϕnsted–Lowry definition, an acid is a proton donor while a base is a proton acceptor. This would mean that in order for an acid–base reaction to occur, both the acid and base must coexist to facilitate proton transfer.

— But what happens to the proton after being accepted by the base? The proton forms a bond with the base. Hence, Gilbert N. Lewis defined an acid as an electron pair acceptor and a base as an electron pair donor (Lewis Theory of acid and base). So, what is the difference between these two definitions? The Brϕnsted–Lowry definition

(Continued)

(*Continued*)

focuses on the process, while the Lewis definition focuses on the outcome, i.e., after the proton has been transferred.

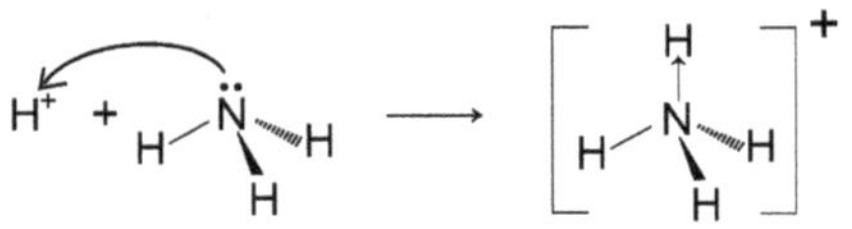

$$H^+ + NH_3 \longrightarrow [NH_4]^+$$

— There are three characteristics of an acid: (1) an acid reacts with metal to give hydrogen gas; (2) an acid reacts with carbonates or hydrogen-carbonates to give carbon dioxide gas; (3) an acid reacts with a base to give salt and water. But not all acids would exhibit all these three characteristics together. For example, water is an acid but it does not react with zinc to give H_2 gas. It also does not react with carbonate to give CO_2 gas. But it does react with sodium metal to give H_2 gas. So whether an acid exhibits each of the above characteristics really depends on the strength of the acid. The stronger the acid, the more likely it would react with a less reactive base or metal.

— A strong acid or base is one that fully dissociates in water, while a weak acid or weak base does not fully dissociate. The fact that a weak acid dissociates to a smaller extent than a strong acid is because the dissociation of a weak acid is energetically less favorable than for a strong acid. A weak acid needs more energy to be supplied in order to dissociate! That is why in Chapter 4, we see that the enthalpy change of neutralization of a weak acid is less exothermic than that of a strong acid. It can even be endothermic in nature, for example, the reaction between HCO_3^- and CH_3COOH is in fact an endothermic value.

(ii) pH.

Explanation:

pH is defined as the negative logarithm of the concentration of H^+ ion or $pH = -\log[H_3O^+]$.

Do you know?

— Similarly, pOH = $-\log[OH^-]$.

— For neutral water, the pH = pOH! If the temperature of water is 25°C, the pH = pOH = 7. This is because water undergoes auto-ionization as follows:

$$H_2O(l) + H_2O(l) \rightleftharpoons H_3O^+(aq) + OH^-(aq).$$

At 25°C, $[H_3O^+] = [OH^-] = 10^{-7}$ mol dm^{-3} and the system is at dynamic equilibrium.

Therefore, we define the *ionic product of water* as:

$$K_w = [H_3O^+][OH^-] = 10^{-14} \text{ mol}^2 \text{ dm}^{-6} \text{ at } 25^\circ\text{C or}$$
$$pK_w = \text{pH} + \text{pOH} = 14 \text{ at } 25^\circ\text{C}.$$

Since the auto-ionization of water is an endothermic process, as the temperature increases, according to Le Chatelier's Principle, the position of the equilibrium shifts to the right to take in the excess heat, and the degree of auto-ionization increases. Thus, although both $[H_3O^+]$ and $[OH^-]$ increase, the water is still neutral in nature as pH = pOH, is still less than 7. In addition, $K_w = [H_3O^+][OH^-] \neq 10^{-14}$ mol^2 dm^{-6}. Hence, pure water is "more acidic" and at the same time, "more alkaline" at a higher temperature. So, in the future, when you use the term 'acidic' or 'basic', you better know what you really intend to mean.

— Now, if we add a strong acid such as HCl(g) into water, it fully dissociates to release more H_3O^+. As such, the 10^{-7} mol dm^{-3} of OH^- ions from the auto-ionization of water will now have a greater chance to encounter a H_3O^+ ion. Hence, the $[OH^-]$ will decrease to a value that is much smaller than 10^{-7} mol dm^{-3}. We say that the auto-ionization of water is being suppressed! But nevertheless, $[H_3O^+][OH^-]$ is still equal to 10^{-14} mol^2 dm^{-6} at 25°C.

— Similarly, when a strong base such as NaOH is added, the increase in $[OH^-]$ would suppress the auto-ionization of water. Thus, $[OH^-]$ would increase while $[H_3O^+]$ would decrease.

— Based on the definition of pH, the higher the $[H^+]$, the lower the pH, the more acidic is the solution, and the higher the $[OH^-]$, the smaller the pOH and the higher the pH value, the more basic is the solution.

(*Continued*)

(*Continued*)

— But we cannot use pH to judge the strength of an acid or base. If we have two strong acids which fully dissociate, the pH of each solution would really depend on the number of moles of each acid molecule that is placed into the system. So you can't say that strong acid A is stronger than strong acid B because it has a lower pH which is actually due to greater number of moles of acid A being placed into the system.

— Now, for a weak acid (HA), it is even more complicated:

$$HA(aq) + H_2O(l) \rightleftharpoons H_3O^+(aq) + A^-(aq).$$

The amount of the weak acid dissociated depends very much on its concentration. According to Le Chatelier's Principle, a higher concentration of a weak acid would cause more acid molecules to dissociate, hence the degree of dissociation (α) increases, which leads to the $[H_3O^+]$ increasing and pH to decrease. But there is a limit to the increase in α and the decrease in pH! This is because as a larger amount of weak acid dissociates, more of the free ions are formed. It would also mean that the backward reaction is more likely to occur. From the above equation, we can see very clearly that in order for HA to function as an acid, we need a base (H_2O) to be present. Imagine if there are too many weak acid molecules but not enough H_2O, can the weak acid dissociate?

— So, we can only use pH or α to compare the strength of two weak acids, provided they are of equal concentration. Second, both must be of similar basicity, meaning the number of moles of OH^- reacting with per mole of acid molecule. For example, both must be monobasic, i.e., one mole of the acid reacts with one mole of OH^- or dibasic or tribasic, etc.

— A more convenient way is to use an equilibrium constant to define the strength of the acid. This is because as long as the system is at equilibrium, an equilibrium constant being a fixed value at a fixed temperature, is independent of the various concentrations of the HA(aq), $H_3O^+(aq)$, and $A^-(aq)$ in the system. This equilibrium constant is known as the acid dissociation constant, $\boldsymbol{K_a}$**:**

$$HA(aq) + H_2O(l) \rightleftharpoons H_3O^+(aq) + A^-(aq), \quad K_a = \frac{[H_3O^+][A^-]}{[HA]}.$$

Like any other equilibrium constant, K_a is temperature dependent! The [HA] term in the K_a expression is the leftover, undissociated HA!

— If it is a weak base, then we have the base dissociation constant, K_b:

$$B(aq) + H_2O(l) \rightleftharpoons BH^+(aq) + OH^-(aq), \quad K_b = \frac{[BH^+][OH^-]}{[B]}$$

If a higher concentration of the weak acid causes more weak acid to dissociate, at the same time, this would also cause the rate of the backward reaction to increase. Then, at a lower concentration, when the weak acid dissociates, there is less chance for the backward reaction to occur because the volume of the system is larger. So, would this cause more $H_3O^+(aq)$ to form and hence lower the pH?

A: There is a fallacy here! Yes, at a lower concentration of the weak acid, you can perceive that dissociation somehow increases because the rate of the backward reaction decreases, so the number of moles of $H_3O^+(aq)$ increases. But do not forget that the volume of the system also increases because a lower concentration of weak acid is equivalent to dilution of a higher concentration of the weak acid. Thus, when you calculate the pH, which is $pH = -\log[H_3O^+] = -\log(\frac{\text{number of moles of } H_3O^+}{\text{volume of solution}})$, the increase in the volume factor is much greater than the increase in the number of moles factor. Hence, overall, the pH increases. Just imagine, if the solution of the weak acid is so much diluted, just like in pure water, would the pH just be about 7?

Q Why didn't we include the $[H_2O]$ into the K_a or K_b expression?

A: The K_a or K_b expression is a "mutated" equilibrium constant, just like K_w. If we consider how K_a came about:

$$HA(aq) + H_2O(l) \rightleftharpoons H_3O^+(aq) + A^-(aq), \quad K_c = \frac{[H_3O^+][A^-]}{[HA][H_2O]}.$$

But since $[H_2O] = \dfrac{\text{Mole}}{\text{Volume}} = \dfrac{\text{Mass}}{\text{Molar mass} \times \text{Volume}} = \dfrac{\rho}{\text{Molar mass}} = \text{constant},$

we have $K_c[H_2O] = K_a = \dfrac{[H_3O^+][A^-]}{[HA]}.$

Can we write the dissociation of a weak acid as: $HA(aq) \rightleftharpoons H^+(aq) + A^-(aq)$?

A: Yes, you can. BUT you have not brought out the important concept that "an acid cannot function as an acid without the presence of a base." Don't you think this equation, $HCl(aq) \rightarrow H^+(aq) + Cl^-(aq)$, is conceptually weird? It is like telling people that the HCl, for no rhyme or reason, somehow automatically breaks up.

Q What is the relationship between HA and A^-?

A: HA and A^- is a conjugate acid–base pair. HA is the conjugate acid of A^-, while A^- is the conjugate base of HA. A conjugate acid–base pair differs by ONLY one H^+.

Q Is H^+ the conjugate acid of OH^-?

A: No! According to this equation, $H_2O(l) + H_2O(l) \rightleftharpoons H_3O^+(aq) + OH^-(aq)$, the conjugate acid of OH^- is H_2O, while the conjugate base of H_3O^+ is also H_2O. So in any acid or base dissociation equation, such as $HA(aq) + H_2O(l) \rightleftharpoons H_3O^+(aq) + A^-(aq)$, we always have a pair of acid–base on the left and another pair of acid–base on the right. This is in line with, "an acid cannot function as an acid without the presence of a base!"

Q What is the K_a and K_b of H_2O?

A: The K_a of $H_2O = \dfrac{[H_3O^+][OH^-]}{[H_2O]}$.

The K_b of $H_2O = \dfrac{[H_3O^+][OH^-]}{[H_2O]}$.

(b) Sulfur dioxide reacts with water to produce sulfuric (IV) acid (H_2SO_3), a weak acid which partially ionizes according to:

$$H_2SO_3(aq) + H_2O(l) \rightleftharpoons H_3O^+(aq) + HSO_3^-(aq).$$

(i) Write an expression in terms of concentrations for the acidity constant (K_a) for sulfuric(IV) acid.

Explanation:

$$H_2SO_3(aq) + H_2O(l) \rightleftharpoons H_3O^+(aq) + HSO_3^-(aq) \quad K_a = \frac{[H_3O^+][HSO_3^-]}{[H_2SO_3]}.$$

Q $H_2SO_3(aq)$ is a dibasic acid, so why is only one of the H^+ ions dissociated?

A: Yes, indeed. One mole of $H_2SO_3(aq)$ requires two moles of OH^-; it is a dibasic acid. But unfortunately, being a weak dibasic acid, the dissociation happens in two steps:

$$H_2SO_3(aq) + H_2O(l) \rightleftharpoons H_3O^+(aq) + HSO_3^-(aq), \qquad K_{a1};$$
$$HSO_3^-(aq) + H_2O(l) \rightleftharpoons H_3O^+(aq) + SO_3^{2-}(aq), \qquad K_{a2}.$$

Q Do K_{a1} and K_{a2} both have the same value?

A: Certainly not! The K_a value for any second dissociation is always smaller than the one before it. This is logical as it becomes more difficult to remove another H^+ ion from a negatively charged species (HSO_3^-) than from a neutral species (H_2SO_3).

Is there any other explanation that can be used to account for why K_{a2} is smaller than K_{a1}?

A: Yes! For *cis*-butenedioic acid (HOOCCH=CHCOOH), also known as maleic acid, the K_{a2} is smaller than K_{a1} because after the first dissociation, the $-COO^-$ group is stabilized through intramolecular hydrogen bonding with the other undissociated –COOH group. The second dissociation is not favored as it would disrupt the intramolecular hydrogen bond that is already formed. In addition, the close proximity of the two $-COO^-$ groups would cause much repulsive force to be generated. Hence, the doubly negatively charged species is highly unfavored.

(ii) The vale of K_a for this reaction at 25°C is 1.5×10^{-2} mol dm^{-3}. Calculate a value for the concentration of dissolved hydrogen ions in a 0.10 mol dm^{-3} solution of sulfuric(IV) acid at 25°C, stating clearly any assumptions made.

Explanation:

Using the I.C.E. table:

	$H_2SO_3(a) + H_2O(1) \rightleftharpoons$	$H_3O^+(aq) +$	$HSO_3^-(aq)$
Initial conc./mol dm^{-3}	0.10	0	0
Change/mol dm^{-3}	$-x$	$+x$	$+x$
At equil.mol dm^{-3}	$0.10-x$	$-x$	$-x$

$$K_a = \frac{[H_3O^+][HSO_3^-]}{[H_2SO_3]} = \frac{(x)(x)}{(0.10-x)} = 1.5\times10^{-2}\ \text{mol dm}^{-3}.$$

Assuming that the dissociation is small, then $0.10 - x \approx 0.10$.
Hence, $x = \sqrt{(1.5 \times 10^{-2} \times 0.10)} = 3.87 \times 10^{-2}$
Therefore, $[H_3O^+] = 3.87 \times 10^{-2}$ mol dm^{-3}.

Do you know?

— The percentage of dissociation for the above $= \frac{0.0387}{0.10}\times100 = 38.7\%$, or the degree of dissociation is 0.387. So, is our assumption that the dissociation is small, valid? Let us try solving this through the quadratic way:

$$x^2 + 1.5\times10^{-2}x - 1.5\times10^{-3} = 0$$

$$\Rightarrow x = \frac{(1.5\times10^{-2}) \pm \sqrt{(1.5\times10^{-2}) - 4(1)(-1.5\times10^{-3})}}{2(1)} = 3.19\times10^{-2}$$

Percentage difference $= \frac{0.0387-0.0319}{0.0319}\times100 = 21.2\%$
Since the percentage difference is greater than 5%, our assumption is not valid.

Q So, when would it be appropriate to make the assumption that the dissociation is small and then do the approximation?

A: A good gauge would be if the K_a or K_b is of a value 10^{-5} or less. Then, when we take the square root, it would at most affect the third or fourth decimal place of our final answer.

(iii) With reference to your answer in part *(b) (ii)*, calculate the pH of a 0.20 mol dm^{-3} solution of sulfuric(IV) acid at 25°C.

Explanation:

$$pH = -\log [H_3O^+] = -\log (3.87 \times 10^{-2}) = 1.41.$$

Do you know?

— There is H_2SO_3 in strawberry jam as SO_2 is used to preserve strawberry. Hence, the SO_2 dissolved in the water to give H_2SO_3, just like CO_2 dissolved to give H_2CO_3.

$$SO_2(g) + H_2O(l) \rightleftharpoons H_2SO_3\ (aq)$$
$$CO_2(g) + H_2O(l) \rightleftharpoons H_2CO_3\ (aq).$$

2. Acids can be differentiated by the number of hydrogen ions that can be liberated from one molecule of the undissociated acid. Hydrochloric acid is a strong monobasic, or monoprotic, acid liberating one hydrogen ion per molecule. Sulfuric(VI) acid is a dibasic, or diprotic, acid, with its ionization in aqueous solution as follows:

$H_2SO_4(l) + H_2O(l) \rightleftharpoons H_3O^+(aq) + HSO_4^-(aq)$, K_a = very large; (I)

$HSO_4^-(aq) + H_2O(l) \rightleftharpoons H_3O^+(aq) + SO_4^{2-}(aq)$, $K_a = 0.01$ mol dm^{-3}. (II)

(K_a values are quoted for 25°C.)

(a) (i) State the hydrogen ion concentration in 0.02 mol dm^{-3} hydrochloric acid, which is a strong acid, and hence find the pH of this solution.

Explanation:

Since HCl(aq) is a strong acid which fully dissociates, $[H_3O^+] = [HCl] = 0.02\ mol\ dm^{-3}$. Therfore, $pH = -\log [H_3O^+] = -\log (0.02) = 1.70$.

Do you know?

— The $[H_2SO_3] = 0.10\ mol\ dm^{-3}$ in Question 1(*b*) and the pH of its solution is 1.41, while the $[HCl] = 0.02\ mol\ dm^{-3}$ and its pH = 1.70. So, what we can see here is that although the concentration of a strong acid is much lower than that of a weak acid, the pH values are about the same. This shows that we cannot use concentration value to judge the strength of an acid.

(ii) State the hydrogen ion concentration of $0.1\ mol\ dm^{-3}$ sulfuric acid arising from the first stage of ionization, (I).

Explanation:

Assuming there is full dissociation for the first step, $[H_3O^+] = 0.1\ mol\ dm^{-3}$.

(iii) A solution of sodium hydrogensulfate, $NaHSO_4$, of concentration $0.1\ mol\ dm^{-3}$, which ionizes according to Eq. (II) above, has a pH of 1.57. Find the hydrogen ion concentration in this solution, and hence state what you would expect the hydrogen ion concentration in $0.1\ mol\ dm^{-3}$ sulfuric acid to be.

Explanation:

$$HSO_4^-(aq) + H_2O(l) \rightleftharpoons H_3O^+(aq) + SO_4^{2-}(aq)$$

Since $pH = -\log [H_3O^+] = 1.57 \Rightarrow [H_3O^+] = 10^{-1.57} = 2.69 \times 10^{-2}\ mol\ dm^{-3}$. The $[H_3O^+]$ in $0.1\ mol\ dm^{-3}$ H_2SO_4 would be $= 0.1 + 2.69 \times 10^{-2} = 0.1269$ $mol\ dm^{-3}$.

Do you know?

— The $[H_3O^+]$ in 0.1 mol dm^{-3} H_2SO_4 is much lower than 0.2 mol dm^{-3} because of the suppression of dissociation in the second step.

(iv) In fact, the pH of 0.1 mol dm^{-3} sulfuric(VI) acid is about 0.98. This indicates a hydrogen ion concentration of 0.105 mol dm^{-3}. Considering that reactions (I) and (II) coexist simultaneously, explain the difference in the hydrogen ion concentrations.

Explanation:

The actual sum of the concentration of the H_3O^+ ions from both dissociation steps is much smaller than the expected value, because of the suppression of the second dissociation by the first dissociation. This is termed the *common ion effect.*

(v) If the K_a for ionization (II) has a value of 0.02 mol dm^{-3} at 80°C, state with reasons, whether the ionization is endothermic or exothermic.

Explanation:

As the temperature increases, K_a also increases from 0.01 to 0.02 mol dm^{-3}. A higher K_a value means a greater degree of dissociation, i.e., position of equilibrium shifts to the right. According to Le Chatelier's Principle, when the temperature increases, the system would respond by removing the excess heat. Since the forward reaction is endothermic in nature, the position of equilibrium would shift to the right.

(vi) Explain the effect of such an increase in temperature on the pH of this solution.

Explanation:

If dissociation increases, then $[H_3O^+]$ increases, which means that the pH should decrease. The solution becomes more acidic.

Do you know?

— As temperature increases, K_w also increases, i.e., $K_w > 10^{-14}$ mol^2 dm^{-6}.

(b) Pure sulfuric(VI) acid has a boiling point of 338°C and it mixes with water in all proportions. Hydrogen chloride boils at −85°C and is extremely soluble in water.

(i) Suggest reasons, in terms of the structure and bonding of H_2SO_4 and HCl, for the large difference between their boiling points.

Explanation:

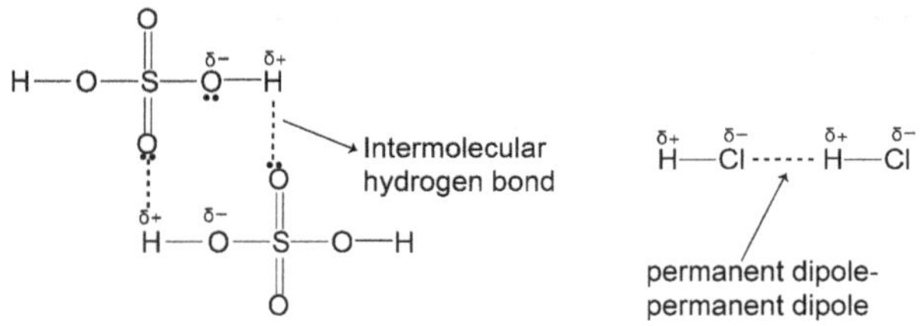

Intermolecular hydrogen bond

permanent dipole-permanent dipole

Both H_2SO_4 and HCl are simple molecular compounds, hence they have weak intermolecular forces between the molecules. The O–H groups present in H_2SO_4 enable the molecules to form strong hydrogen bonds among themselves. On the other hand, for HCl, being a polar molecule, it can only form weaker permanent dipole–permanent dipole interaction. This explains why the boiling point of H_2SO_4 is higher than that of HCl.

If the H atom of the HCl interacts with the lone pair of electrons of the O atom of H_2SO_4, is this considered hydrogen bond?

A: By definition, no! It is still permanent dipole–permanent dipole interaction.

(ii) Suggest why both compounds are so soluble in water.

Explanation:

Both are so soluble in water because both dissociate to give free ions, such as H_3O^+, HSO_4^-, SO_4^{2-}, and Cl^-. These ions can form strong ion–dipole interaction with the water molecules. This process releases much energy to help in the solubility and the dissociation.

Do you know?

— In order for a solute to dissolve well in a solvent, the solute–solvent interaction must be of equivalent strength to the solute–solute and solvent–solvent interaction. This is the explanation behind the maxim 'like dissolves like'. With similar solute–solvent interaction, there would then be sufficient amount of energy released to overcome the solute–solute and solvent–solvent interactions. This is why it is important to first know the type of intermolecular forces that is individually present in the solute and the solvent molecules, before predicting whether they would dissolve each other.

— As for dissolution of ionic compounds, it is the balance between the exothermic hydration enthalpies of the cation and anion and the energy needed to break up the ionic lattice. The energy that is needed to break up the ionic lattice is numerically equivalent to the lattice energy. If the sum of the numerical value for the enthalpies of hydration of the cation and anion is greater than the numerical value of the lattice energy, then the ionic solid is likely to dissolve.

(iii) The K_{sp} of magnesium sulfate is smaller than that of barium sulfate; use the following data to suggest why their solubilities differ considerably: $\Delta H_{hydration}$ of $Mg^{2+} = -1920\ kJ\ mol^{-1}$, $Ba^{2+} = -1360\ kJ\ mol^{-1}$.

Explanation:

A smaller K_{sp} value means a lower solubility here. The energy needed to break up the ionic lattice has to be supplemented by the hydration enthalpies. Since the enthalpy of hydration for the $SO_4^{2-}(aq)$ ion is the same for both sulfates, then the more exothermic $\Delta H_{hydration}$ of Mg^{2+} should make $MgSO_4$ more soluble. But since the solubility of $MgSO_4$ is actually lower than $BaSO_4$, it can only mean that the energy needed to break up the ionic lattice for $MgSO_4$ is more endothermic than that for $BaSO_4$, so much so that the more exothermic $\Delta H_{hydration}$ of Mg^{2+} cannot compensate for it.

So, we can actually compare the solubilities of ionic compounds just by inspecting their K_{sp} values?

A: You can gauge the strength of an acid or base just by looking at its K_a or K_b value. But you cannot do the same for the K_{sp} values of the ionic compounds. Let us explain.

For an ionic compound, AB(s):

$$AB(s) \rightleftharpoons A^+(aq) + B^-(aq), \quad K_{sp} = [A^+][B^-],$$

the solubility, s, is $s^2 = [A^+][B^-] = K_{sp} \Rightarrow s = \sqrt{Ksp}$.

For an ionic compound, $A_nB(s)$:

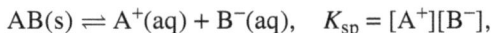

$$A_nB(s) \rightleftharpoons nA^+(aq) + B^{n-}(aq), \quad K_{sp} = [A^+]^n[B^{n-}] \text{ where } \frac{[A^+]}{[B^{n-}]} = \frac{n}{1},$$

the solubility, s, is $K_{sp} = [A^+]^n[\ B^{n-}] = (ns)^n.s = s^{(n+1)}.(n)^n \Rightarrow s = \sqrt[n+1]{\frac{Ksp}{n^n}}$.

For an ionic compound, $AB_n(s)$:

$$AB_n(s) \rightleftharpoons A^{n+}(aq) + nB^-(aq) \quad K_{sp} = [A^{n+}][\ B^-]^n \text{ where } \frac{[A^{n+}]}{[B^-]} = \frac{1}{n},$$

the solubility, s, is $K_{sp} = [A^{n+}][B^-]^n = s.\ (ns)^n = s^{(n+1)}.(n)^n \Rightarrow s = \sqrt[n+1]{\frac{Ksp}{n^n}}$.

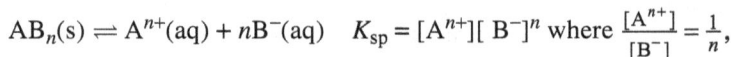

So, if you compare the solubility of AB(s) and A_nB(s) just by looking at their K_{sp} values, you would reach the wrong conclusion. Thus, comparison of K_{sp} values to decide the solubility is ONLY valid provided the sum of the powers of the concentration of all the ions in the K_{sp} expression are the same, for example, A_nB(s) and AB_n(s).

3. A monobasic carboxylic acid **W** contains 40.0% carbon, 6.70% hydrogen, and 53.3% oxygen by mass. When 10.0 cm^3 of an aqueous solution of **W**, with concentration 4.65 g dm^{-3}, is titrated against 0.050 mol dm^{-3} sodium hydroxide, the following pH readings are obtained.

Volume of NaOH (cm^3)	0.0	2.5	5.0	7.5	10.0	14.0	15.0	16.0	17.5	20.0	22.5
pH	2.5	3.2	3.5	3.8	4.1	4.7	5.2	9.1	11.5	11.8	12.0

(a) Plot a graph of pH against volume of NaOH used in the titration. Use the graph to determine the endpoint of the titration. Hence, calculate the relative molecular mass of **W**.

Explanation:

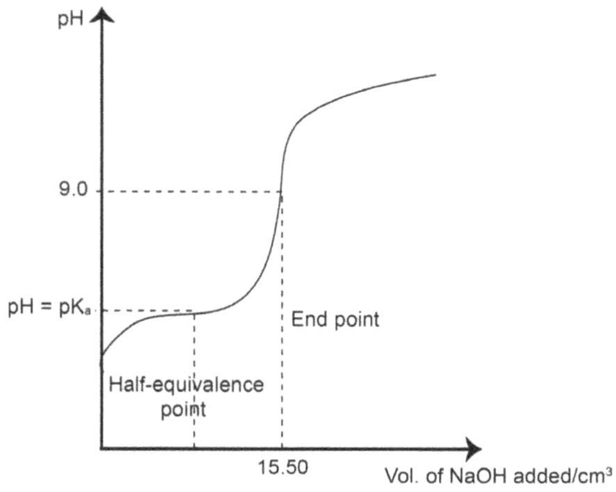

From the graph, volume of NaOH needed = 15.50 cm^3.
Amount of NaOH needed = 15.50/1000 × 0.05 = 7.75 × 10^{-4} mol.
Amount of **W** in 10 cm^3 = Amount of NaOH needed = 7.75 × 10^{-4} mol.

Concentration of **W** = $7.75 \times 10^{-4}/0.010 = 0.0775\,\text{mol dm}^{-3}$.
Amount of **W** in 4.65 g = $\frac{4.65}{\text{Molar mass}} = 0.0775$.
Therefore, molar mass of **W** = $\frac{4.65}{0.0775} = 60.0\,\text{g mol}^{-1}$.

Do you know?

— A carboxylic acid is a weak organic acid that contains the −COOH functional group.

— The pH of the end point for a weak acid–strong base titration is not neutral! In fact, it would be greater than 7. Why? A weak acid is weak, meaning the weak acid molecule, HA, does not "like" to dissociate. This indicates that the conjugate base of the weak acid, i.e., A^-, would have a high affinity for a H^+ ion. Hence, what happens is that at the endpoint of titration, the conjugate base undergoes basic hydrolysis as shown below:

$$A^-(aq) + H_2O \rightleftharpoons HA(aq) + OH^-(aq), \quad K_b = \frac{[HA][OH^-]}{[A^-]}.$$

And if we take this K_b and multiply it by the K_a of HA, which is $K_a = \frac{[H_3O^+][A^-]}{[HA]}$, we get

$$K_a \times K_b = \frac{[H_3O^+][A^-]}{[HA]} \times \frac{[HA][OH^-]}{[A^-]} = [H_3O^+][OH^-] = K_w.$$

Or, $pK_a + pK_b = pH + pOH = pK_w = 14$!
Thus, if we are given the K_a of a weak acid HA, we can find the K_b of its conjugate base using the above expression, and vice versa!

— The pH value at the half-equivalence point is useful as it gives us the K_a value. How? Let us explain:

From $K_a = \frac{[H_3O^+][A^-]}{[HA]} \Rightarrow -\log K_a = -\log[H_3O^+] - \log\frac{[A^-]}{[HA]}$

$$\Rightarrow pK_a = pH - \log\frac{[A^-]}{[HA]}$$

$$\Rightarrow pH = pK_a + \log\frac{[A^-]}{[HA]}$$

(Henderson–Hasselbach Equation).

(Continued)

(Continued)

This equation is useful as it allows us to find the pH of a solution that contains a mixture of the weak acid, HA, and its conjugate base, A^-, especially along the titration curve. Now, at the mid-point of titration, $[HA] = [A^-]$, hence $pH = pK_a$!

— Did you notice that the pH change near the mid-point of titration is very gradual, and not as drastic as the one for a strong acid–strong base titration that is shown below?

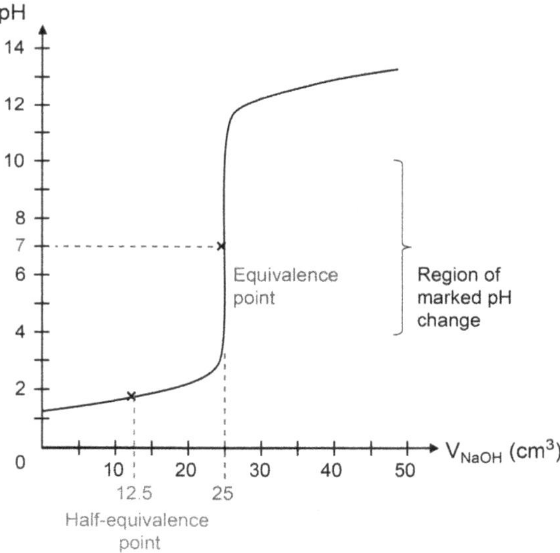

This is because as the HA is being neutralized, it generates the conjugate base, A^-. A mixture of HA and A^- constitutes a buffer solution, which maintains a relatively constant pH when a small amount of acid or base is added.

Q How does a buffer keep the pH constant?

A: A buffer consists of a mixture of HA and A^-. The way we should explain how a buffer works is as follows:

When a small amount of H_3O^+ is added: $H_3O^+ + A^- \rightarrow HA + H_2O$.

When a small amount of OH^- is added: $OH^- + HA \rightarrow A^- + H_2O$.

The presence of large concentrations of HA and A^- ensures that the pH changes negligibly.

If a buffer consists of a mixture of HA and A^-, how would we know the pH of this buffer solution? That is, whether it is an acidic or alkaline buffer?

A: Simple! Look at the K_a value of HA and the K_b value of A^-. If the K_a value of HA is greater than the K_b value of A^-, then the solution would be overall an acidic buffer. Likewise, if K_a value of HA is smaller than the K_b value of A^-, then it would mean that the basic hydrolysis of A^- is more pronounced than the dissociation of HA. Hence, an alkaline solution would result!

Since the pH of the end point of a weak acid–strong base titration is above pH 7, what should be the suitable indicator to use for detecting the end point?

A: You must use an indicator where the working range coincides with the region of drastic pH change in the titration curve. The following shows the working range and the respective color changes for some common indicators:

	Methyl orange	Screened methyl orange	Phenolphthalein
K_{In}	3.7	3.7	9.3
pH range/working range	3.1–4.4	3.1–4.4	8.2–10
color in "acid" solution	red	violet	colorless
color in "alkaline" solution	yellow	green	pink

Thus, for a weak acid–strong base titration, phenolphthalein is a good choice while for a weak base–strong acid titration, both methyl orange and screened methyl orange can be used.

Why is the pH of the end point for a weak base–strong acid titration in the acidic pH range?

A: If we consider the weak base, NH_3, it hydrolyzes in water as follows:

$$NH_3(aq) + H_2O(l) \rightleftharpoons NH_4^+(aq) + OH^-(aq).$$

At the end point of the titration, the conjugate acid, $NH_4^+(aq)$ would undergo the following dissociation, resulting in the formation of an acidic solution:

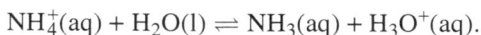

$$NH_4^+(aq) + H_2O(l) \rightleftharpoons NH_3(aq) + H_3O^+(aq).$$

Q What is the difference between 'hydrolysis' and 'dissociation'?

A: Hydrolysis refers to the reaction where a H_2O molecule is being split up, like:

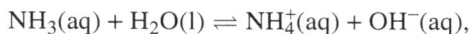

$$NH_3(aq) + H_2O(l) \rightleftharpoons NH_4^+(aq) + OH^-(aq),$$

whereas dissociation refers to a species being broken up, like:

$$NH_4^+(aq) + H_2O(l) \rightleftharpoons NH_3(aq) + H_3O^+(aq).$$

So, though both are acid–base in nature here, they are actually quite different!

(b) Calculate the value of K_a for **W** and state its units.

Explanation:

At the half-equivalence point, pH = pK_a = 4.0, hence $K_a = 10^{-4.0} = 1.0 \times 10^{-4}\,\text{mol dm}^{-3}$.

(c) Calculate the molecular formula of **W**.

Explanation:

To calculate the empirical formula:

	C	H	O
Percentage by mass	40	6.7	53.3
No. of mole in 100 g	3.33	6.7	3.33
Mole ratio	1	2	1

The empirical formula is CH_2O.
Since, n(E.F.) = M.F. $\Rightarrow n(CH_2O) = 60.0$, i.e., $n \times (12.0 + 2.0 + 16.0) = 60.0$
Therefore, $n = 2$. Hence, the molecular formula of **W** is $C_2H_4O_2$.

Do you know?

— The titration method can be used to determine the molecular formula, as shown in this question.

4. (a) When ethanoic acid is dissolved in water, the following equilibium is established:

$$CH_3CO_2H + H_2O \rightleftharpoons CH_3CO_2^- + H_3O^+. \quad \text{(I)}$$

When hydrogen chloride dissolves in ethanoic acid, the equilibrium established is:

$$CH_3CO_2H + HCl \rightleftharpoons CH_3CO_2H_2^+ + Cl^-. \quad \text{(II)}$$

What is the role of the ethanoic acid in:

(i) equilibrium (I), and

Explanation:

Since CH_3CO_2H loses a H^+ to become $CH_3CO_2^-$, it is a proton donor. Based on the Brϕnsted–Lowry definition, it is acting as an acid.

(ii) equilibrium (II).

Explanation:

Since CH_3CO_2H gains a H^+ to become $CH_3CO_2H_2^+$, it is a proton acceptor. Based on the Brϕnsted–Lowry definition, it is acting as a base.

Do you know?

— The role of a molecule that is acting as an acid or a base is really very relative. It depends on what other reactant is present with it. Take for instance, a lot of students think that ammonia can only act as a base, but ammonia is actually acidic enough to react with sodium metal to give hydrogen gas:

$$Na(s) + NH_3(l) \rightarrow NaNH_2(s) + \frac{1}{2} H_2(g).$$

Many also think that acidic hydrogen must come from a $-OH$ group, HCl, HBr, or HI. But $H-CN$ is a weak acid:

$$HCN(aq) + H_2O(l) \rightleftharpoons CN^-(aq) + H_3O^+(aq).$$

In a nutshell, as long as a species contain a lone pair of electrons that potentially can be donated, then this species potentially can act as a base. Vice versa, a species that contains a H atom that potentially can be denoted as a H^+, it potentially can act as an acid!

(b) What is the relationship between the species $CH_3CO_2H_2^+$ and CH_3CO_2H?

Explanation:

Since $CH_3CO_2H_2^+$ and CH_3CO_2H only differ by a H^+, they are a conjugate acid–base pair.

(c) The value of K_a for ethanoic acid at 25°C is 1.80×10^{-5} mol dm^{-3} and for methanoic acid, HCO_2H, it is 1.56×10^{-4} mol dm^{-3}.

(i) Give the K_a expression for CH_3CO_2H.

Explanation:

$CH_3CO_2H(aq) + H_2O(l) \rightleftharpoons CH_3CO_2^-(aq) + H_3O^+(aq)$, $K_a = \frac{[H_3O^+][CH_3CO_2^-]}{[CH_3COOH]}$.

(ii) Hence, calculate the pH of a 0.100 mol dm^{-3} solution of CH_3CO_2H at 25°C.

Explanation:

Using the I.C.E. table: $CH_3CO_2H(aq) + H_2O(l) \rightleftharpoons H_3O^+(aq) + CH_3CO_2^-(aq)$

	CH_3CO_2H	H_3O^+	$CH_3CO_2^-$
Initial conc./mol dm^{-3}	0.10	0	0
Change/mol dm^{-3}	$-x$	$+x$	$+x$
At equil/mol dm^{-3}	$0.10-x$	$+x$	$+x$

$$K_a = \frac{[H_3O^+][CH_3CO_2^-]}{[CH_3COOH]} = \frac{(x)(x)}{(0.10-x)} = 1.80\times10^{-5}\,\text{mol dm}^{-3}.$$

Assuming that the dissociation is small, then $0.10 - x \approx 0.10$.
Hence, $x = \sqrt{(1.80 \times 10^{-5} \times 0.10)} = 1.34 \times 10^{-3}$.
Therefore, $[H_3O^+] = 1.34 \times 10^{-3}$ mol dm^{-3}.
$\Rightarrow$ pH $= -\log(1.34 \times 10^{-3}) = 2.87$.

(d) The pH of a 0.050 mol dm^{-3} solution of methanoic acid is 2.55. Using this, together with the data in *(c)* and your answer to *(c) (ii)*:

(i) state which of the two acids is the stronger acid; and

Explanation:

Since the K_a of ethanoic acid at 25°C is 1.80×10^{-5} mol dm^{-3} and is smaller than the K_a for methanoic acid, HCO_2H, at 1.56×10^{-4} mol dm^{-3}, ethanoic acid is a weaker acid than methanoic acid.

(ii) comment on the relative pH values of the two acids.

Explanation:

[Ethanoic acid] = 0.10 mol dm^{-3} and pH = 2.87, while [methanoic acid] = 0.05 mol dm^{-3} and pH = 2.55. This shows that although methanoic acid is a stronger acid than ethanoic acid, but because the concentration of ethanoic acid is higher, there is a greater degree of dissociation. Hence, the pH value of ethanoic acid is near to that of the methanoic acid, although methanoic acid is stronger than ethanoic acid.

Do you know?

— This is a good example to show that the concentration of an acid can affect the degree of dissociation. Hence, it is not good enough to judge the strength of two weak acids just by looking at their pH values, especially if their concentrations are not the same.

(e) (i) Sketch a graph showing how the pH changes during the titration of 20.0 cm^3 of a 0.100 mol dm^{-3} solution of methanoic acid with 0.050 mol dm^{-3} sodium hydroxide solution.

Explanation:

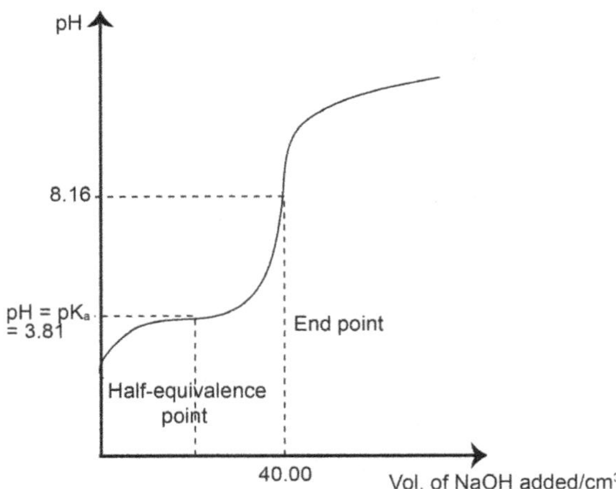

(ii) Select from below a suitable indicator for this titration. Give a brief reason for your choice based on the curve drawn in *(e) (i)*.

Indicator	pH Range
Bromocresol green	3.5–5.4
Bromothymol blue	6.0–7.6
Phenol red	6.8–8.4

Explanation:

Phenol red is a suitable indicator for the titration as the working range (6.8–8.4) coincides with the drastic pH change at the end point, which is around pH = 8.16.

(iii) There is no suitable indicator for the titration of methanoic acid with ammonia. Why is this?

Explanation:

At the end point of the titration, both $HCOO^-$ and NH_4^+ are generated. $HCOO^-$ would undergo basic hydrolysis:

$$HCOO^-(aq) + H_2O(l) \rightleftharpoons OH^-(aq) + HCOOH(aq).$$

NH_4^+ will dissociate to give:

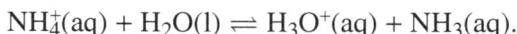

$$NH_4^+(aq) + H_2O(l) \rightleftharpoons H_3O^+(aq) + NH_3(aq).$$

A mixture of $OH^-(aq)$ and $H_3O^+(aq)$ would cause the end point pH to be very undefined. Hence, without a drastic change in pH, it is very difficult to find a suitable indicator.

How will we know whether the end point pH for such a titration would be acidic or alkaline in nature?

A: Look at the K_a of NH_4^+ versus the K_b of $HCOO^-$. If the K_a is larger than the K_b, then we would have an acidic solution. If it is the other way round, then it would be an alkaline solution.

5. A buffer solution of pH 3.87 contains 7.40 g dm^{-3} of propanoic acid ($CH_3CH_2CO_2H$) together with a quantity of sodium propanoate ($CH_3CH_2CO_2Na$). K_a for propanoic acid = 1.35×10^{-5} mol dm^{-3} at 298 K.

   (a) Explain what a buffer solution is and how this particular solution achieves its buffer function.

Explanation:

A buffer solution is one that resists a change in pH when a small amount of acid or base is added. It consists of a mixture of $CH_3CH_2CO_2H$ and $CH_3CH_2CO_2Na$.

When a small amount of H_3O^+ is added:

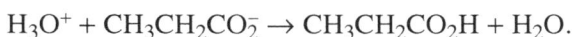

$$H_3O^+ + CH_3CH_2CO_2^- \rightarrow CH_3CH_2CO_2H + H_2O.$$

When a small amount of OH^- is added:

$$CH_3CH_2CO_2H + OH^- \rightarrow CH_3CH_2CO_2^- + H_2O.$$

The presence of large concentrations of $CH_3CH_2CO_2H$ and $CH_3CH_2CO_2^-$ ensure that the pH changes negligibly.

(b) Calculate the concentration (in g dm^{-3}) of sodium propanoate in the solution, stating any assumptions made.

Explanation:

Molar mass of $CH_3CH_2CO_2H$ = 15 + 14 + 44 + 1 = 74 g mol^{-1}.

Concentration of $CH_3CH_2CO_2H$ = 7.40/74 = 0.1 mol dm^{-3}.

Assuming that after mixing, the concentrations of $CH_3CH_2CO_2H$ and $CH_3CH_2CO_2Na$ remain the same.

Using the Henderson–Hasselbach Equation,

$$pH = pK_a + \log\frac{[\text{conjugate base}]}{[\text{conjugate acid}]}$$

$$\Rightarrow 3.87 = -\log(1.35\times10^{-5}) + \log\frac{[CH_3CH_2CO_2Na]}{0.1}$$

$$\Rightarrow [CH_3CH_2CO_2Na] = 0.010 \text{ mol dm}^{-3}.$$

Molar mass of $CH_3CH_2CO_2Na$ = 15 + 14 + 44 + 23 = 96 g mol^{-1}.

Concentration of $CH_3CH_2CO_2Na$ = 0.010 × 96 = 0.961 g dm^{-3}.

Q When the $CH_3CH_2CO_2Na$ is added into the system, wouldn't it undergo hydrolysis to give $CH_3CH_2CO_2H$? Thus the concentration of $CH_3CH_2CO_2H$ should be more than 0.1 mol dm^{-3}?

A: Yes, when $CH_3CH_2CO_2^-$ is added into the system, it would undergo basic hydrolysis to generate $CH_3CH_2CO_2H$:

$$CH_3CH_2CO_2^- + H_2O \rightleftharpoons OH^- + CH_3CH_2CO_2H.$$

At the same time, the $CH_3CH_2CO_2H$ that is added would also dissociate to give $CH_3CH_2CO_2^-$. With these competing equilibria existing, we assume that the addition of $CH_3CH_2CO_2^-$ would suppress the dissociation of $CH_3CH_2CO_2H$. Similarly, the addition of $CH_3CH_2CO_2H$ would also suppress the hydrolysis of $CH_3CH_2CO_2^-$ through the *common ion effect*. Hence, the number of moles of the components that make up the buffer solution are the same as their original number of moles' values.

(c) If the sodium propanoate were to be replaced by anhydrous magnesium propanoate, calculate the concentration of magnesium propanoate (in g dm^{-3}) required to give a buffer of the same pH.

Explanation:

Concentration of $CH_3CH_2CO_2Na = 0.010\,mol\,dm^{-3}$.
Amount of $(CH_3CH_2CO_2)_2Mg$ needed = ½ × 0.010 = 0.005 mol.
Molar mass of $(CH_3CH_2CO_2)_2Mg = (15 + 12 + 44) \times 2 + 24.3 = 166.3\,g\,mol^{-1}$.
Concentration of $(CH_3CH_2CO_2)_2Mg = 0.005 \times 166.3 = 0.832\,g\,dm^{-3}$.

6. Chromium (III) hydroxide is a compound commonly found on the surface of chromed metal that acts as a protective layer. The layer of $Cr(OH)_3$ is usually formed through anodic oxidation in an alkaline solution. The solubility of $Cr(OH)_3$ in pure water is $1.3 \times 10^{-8}\,mol\,dm^{-3}$.
   (a) Determine the K_{sp} of $Cr(OH)_3$.

Explanation:

$$Cr(OH)_3(s) \rightleftharpoons Cr^{3+}(aq) + 3OH^-(aq), \qquad K_{sp} = [Cr^{3+}][OH^-]^3.$$

$[Cr^{3+}] = 1.3 \times 10^{-8}\,mol\,dm^{-3}$.
$[OH^-] = 3 \times 1.3 \times 10^{-8} = 3.9 \times 10^{-8}\,mol\,dm^{-3}$.
Therefore, $K_{sp} = [Cr^{3+}][OH^-]^3 = (1.3 \times 10^{-8})(3.9 \times 10^{-8})^3 = 7.71 \times 10^{-31}\,mol^4\,dm^{-12}$

Do you know?

— We can make use of the K_{sp} value to predict whether a precipitate would be observed:
- If ionic product $>K_{sp}$, precipitation would occur until ionic product $= K_{sp}$ (the solution becomes saturated).
- If ionic product $<K_{sp}$, precipitation would not occur.

The ionic product simply has the same formula as the K_{sp} expression. It depicts the product of the instantaneous concentration of each of the ions present in the system.

Q Why would precipitation occur when ionic product $>K_{sp}$?

A: When a partially soluble compound cannot dissolve any further, it would mean that the precipitation rate is equal to the solution rate. We have what we call a saturated solution. When we mix two ions together, if the rate of precipitation is greater than the rate of dissolving, then precipitation would occur until both the forward rate and backward rate are the same. If you mix two ions together, and the precipitation rate is lower than the solution rate, you won't see any precipitate formed.

Q Why does SnS and CuS form in acidic H_2S solution, while FeS needs alkaline H_2S solution?

A: H_2S is a weak acid:

$$H_2S(aq) + H_2O(l) \rightleftharpoons H_3O^+(aq) + HS^-(aq)$$
$$HS^-(aq) + H_2O(l) \rightleftharpoons H_3O^+(aq) + S^{2-}(aq).$$

Due to the small K_{sp} values of the SnS and CuS, in an acidic solution of H_2S, the ionic product easily exceeds the K_{sp} value. Hence,precipitation occurs. But for FeS, due to a larger K_{sp} value, OH^- needs to be added to generate more S^{2-} by shifting the above equilibria to the right, so that the ionic product for the formation of FeS can exceed its K_{sp} value.

(b) Sketch a diagram to show the changes in the concentration of the ions in pure water as $Cr(OH)_3$ dissolves.

Explanation:

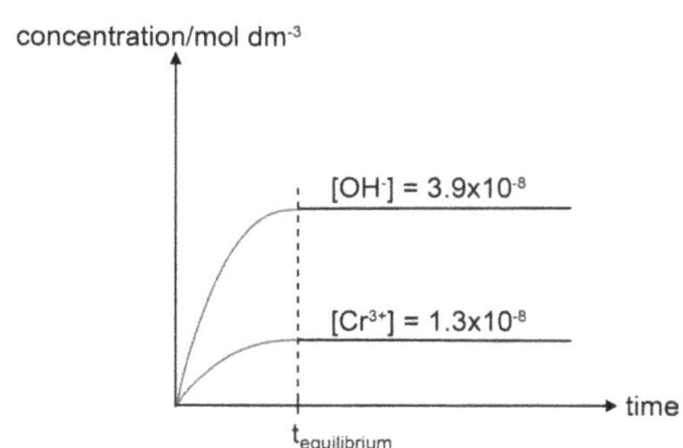

$[Cr^{3+}]:[OH^-] = 1:3$ in pure water!

(c) Determine the solubility of $Cr(OH)_3$ in 0.10 mol dm^{-3} $CrCl_3$.

Explanation:

Let the solubility of $Cr(OH)_3$ be s mol dm^{-3}.
Using the I.C.E. table:

	$Cr(OH)_3(s)$	$\rightleftharpoons$	$Cr^{3+}(aq)$	+	$3OH^-(aq)$
Initial conc/mol dm^{-3}			0.10		0
Change/mol dm^{-3}			$+s$		$+3s$
At equil. mol dm^{-3}			$(0.10 + s)$		$3s$

$$K_{sp} = [Cr^{3+}][OH^-]^3 = (0.10 + s)(3s)^3 = 7.71 \times 10^{-31}\ \text{mol}^4\,\text{dm}^{-12}$$

Due to common ion effect, assume $s << 0.10$, then $0.10 + s \approx 0.10$ mol dm^{-3}.
Therefore, $s = 6.59 \times 10^{-11}$ mol dm^{-3}.
Total $[Cr^{3+}] = 0.10$ mol dm^{-3}.
$[OH^-] = 3 \times 6.59 \times 10^{-11} = 1.98 \times 10^{-10}$ mol dm^{-3}.

Do you know?

— When a common ion is present in the system, the solubility of the compound decreases. For example, in pure water, $[Cr^{3+}]:[OH^-] = 1:3$. But in the presence of 0.10 mol dm^{-3} $CrCl_3$, $[Cr^{3+}]:[OH^-] = 0.10:1.98\times10^{-10}$ $= 5.06\times10^8:1$. So, take note that in the presence of a common ion, the ratio of the species in the system is no longer equal to the ratio of their stoichiometric coefficients.

(d) Indicate the changes in solubility of $Cr(OH)_3$ when 0.10 mol of solid $CrCl_3$ is added to 1 dm^3 of a saturated solution of $Cr(OH)_3$ on the same diagram in part (b).

Explanation:

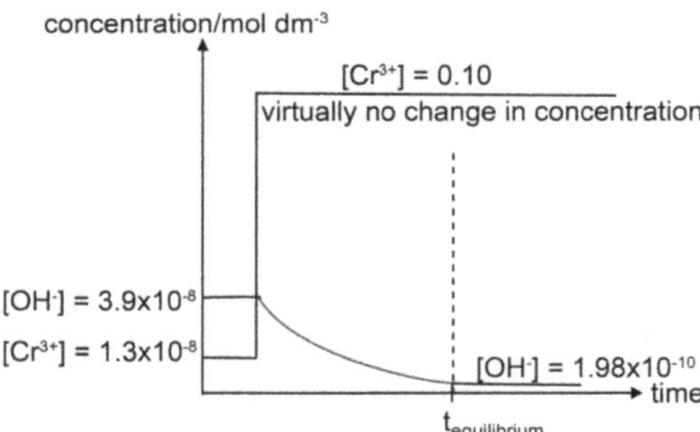

(e) When $Cr(OH)_3$ dissolves in water, it is found that $[Cr^{3+}]:[OH^-]$ is actually greater than 1:3. Give a reasonable explanation for this observation.

Explanation:

When $Cr(OH)_3$ dissolves in water, Cr^{3+} actually forms the $[Cr(H_2O)_6]^{3+}$ complex. But due to the high charge density of the Cr^{3+} ion, the electron cloud of the H_2O molecule is being polarized to an extent whereby the O−H bond is much weakened, causing the $[Cr(H_2O)_6]^{3+}$ complex to undergo appreciable acidic hydrolysis:

$$[Cr(H_2O)_6]^{3+} + H_2O \rightleftharpoons H_3O^+ + [Cr(H_2O)_5(OH)]^{2+}.$$

As a result, the concentration of $[Cr(H_2O)_6]^{3+}$ decreases. According to Le Chatelier's Principle, the position of the following equilibrium shifts to the right, causing more $Cr(OH)_3$ to dissolve.

$$Cr(OH)_3(s) \rightleftharpoons Cr^{3+}(aq) + 3OH^-(aq).$$

Hence, $[Cr^{3+}]$:$[OH^-]$ becomes greater than 1:3 or in another perspective, the H_3O^+ generated from the acidic hydrolysis neutralizes the OH^- ion, causing the ratio to be greater than 1:3..

Do you know?

— Other than Cr^{3+}, other triply charged cations like Al^{3+} and Fe^{3+} can also undergo appreciable hydrolysis to give a highly acidic solution:

$$[Al(H_2O)_6]^{3+} + H_2O \rightleftharpoons H_3O^+ + [Al(H_2O)_5(OH)]^{2+}$$
$$[Fe(H_2O)_6]^{3+} + H_2O \rightleftharpoons H_3O^+ + [Fe(H_2O)_5(OH)]^{2+}.$$

The hydrolysis is so appreciable that when carbonate or hydrogencarbonate is added into the solution, CO_2 gas evolves. This would mean that $Fe_2(CO_3)_3$, $Al_2(CO_3)_3$, and $Cr_2(CO_3)_3$ cannot be synthesized!

— In addition, further hydrolysis of the complex can take place:

$$[Al(H_2O)_5(OH)]^{2+} + H_2O \rightleftharpoons H_3O^+ + [Al(H_2O)_4(OH)_2]^+$$
$$[Al(H_2O)_4(OH)_2]^+ + H_2O \rightleftharpoons H_3O^+ + [Al(H_2O)_3(OH)_3](s).$$

But obviously, the subsequent hydrolysis steps are not as significant as the first step due to common ion effect created by the H_3O^+ from the first hydrolysis step. Now, when NaOH(aq) is added, a white precipitate of $Al(OH)_3$ is formed. The formation of the precipitate can be perceived as a shift of equilibrium, rather than involving the collision of one Al^{3+} and three OH^- ions simultaneously, which is statistically improbable.

(f) When a solution of H_2O_2 is added to a saturated solution containing the undissolved $Cr(OH)_3$, a yellow solution containing CrO_4^{2-} is formed. All the insoluble $Cr(OH)_3$ becomes soluble. With the aid of appropriate equations, explain the observation.

Explanation:

When H_2O_2 is added, the following redox reaction takes place:

$$3H_2O_2 + 2Cr^{3+} + 2H_2O \rightarrow 2CrO_4^{2-} + 10H^+.$$

This decreases the $[Cr^{3+}]$, therefore according to Le Chatelier's Principle, the position of equilibrium for the following reaction shifts to the right:

$$Cr(OH)_3(s) \rightleftharpoons Cr^{3+}(aq) + 3OH^-(aq).$$

This causes all the $Cr(OH)_3$ to dissolve.

Do you know?

— What are the factors affecting solubility?
- Common ion effect: decreases solubility.
- Removal of cation:

Examples:

- ○ AgCl dissolves when $NH_3(aq)$ is added because the Ag^+ ion from the equilibrium, $AgCl(s) \rightleftharpoons Ag^+ + Cl^-$, is removed through the formation of $[Ag(NH_3)_2]^+(aq)$.
- ○ The solubility of $Cu(OH)_2$ with the addition of $NH_3(aq)$ is due to the removal of the Cu^{2+} ion through the formation of the $[Cu(NH_3)_4(H_2O)_2]^{2+}$ complex.
- ○ The solubility of $Zn(OH)_2$ with the addition of NaOH(aq) is due to the removal of the Zn^{2+} ion through the formation of the $[Zn(OH)_4]^{2-}$ complex.
- ○ The solubility of $PbCl_2$ with the addition of concentrated HCl is due to the removal of the Pb^{2+} ion through the formation of the $[PbCl_4]^{2-}$ complex.

(*Continued*)

(*Continued*)

- Removal of anion:

  Examples:

  - $BaCrO_4$ dissolves in dilute HNO_3 because the CrO_4^{2-} from the equilibrium, $BaCrO_4(s) \rightleftharpoons Ba^{2+} + CrO_4^{2-}$, is removed through the formation of $Cr_2O_7^{2-}(aq)$.
  - The solubility of PbI_2 with the addition of H_2O_2 is due to the removal of the I^- ion through the formation of the I_2.
  - The solubility of $Ag_2C_2O_4$ with the addition of dilute HNO_3 is due to the removal of the $C_2O_4^{2-}$ ion through the formation of the $H_2C_2O_4$.
  - The solubility of $Mg(OH)_2$ with the addition of solid NH_4Cl is due to the removal of the OH^- ion through the reaction of OH^- with the acidic NH_4^+.
  - The solubility of $Pb_3(PO_4)_2$ with the addition of dilute HNO_3 is due to the removal of the PO_4^{3-} ion through the formation of HPO_4^{2-}.

(g) $Al(OH)_3$ is another hydroxide that is insoluble in water. Briefly describe how you would determine which compound, $Al(OH)_3$ or $Cr(OH)_3$, is more soluble in water.

Explanation:

Put some $Al(OH)_3$ and $Cr(OH)_3$ separately into the same volume of water. After sometime, when none of the solid can be further dissolved, use a pH probe to measure the pH of each solution. The solution that gives a higher pH value must be the more soluble compound.

Alternatively, dissolve as much of the compound in water, then filter the mixture and collect the filtrate. Pipette a fixed volume of the filtrate and titrate with standard HCl solution. The filtrate that needs more HCl for neutralization must contain more OH^-, hence indicating a higher solubility of the compound.

Alternatively, pipette a fixed volume of the filtrate and evaporate to dryness. Then measure the mass of the residue.

(h) The ΔG^{θ} of the dissolution of $Cr(OH)_3$ can be determined from the following equation:

$$\Delta G^{\theta} = -RT \ln K_{sp}.$$

With reference to ΔG^{θ}, explain why $Cr(OH)_3$ is sparingly soluble in water at 25°C.

Explanation:

$\Delta G^{\theta} = -RT \ln K_{sp} = -(8.314)(298) \ln (7.71 \times 10^{-31}) = +171.8\,\text{kJ mol}^{-1}$. Since the ΔG^{θ} for the solubility of $Cr(OH)_3$ is endergonic, the solubility process is thermodynamically non-spontaneous under standard conditions.

Do you know?

— All equilibrium constants are related to the Gibbs Free Energy change, ΔG^{θ}, via:

$$\Delta G^{\theta} = -RT \ln K \text{ where } K \text{ is the equilibrium constant.}$$

And we can calculate the ΔG^{θ} for the equilibrium and predict its spontaneity.

(i) The solubility of $Cr(OH)_3$ increases when the temperature is increased. With reference to ΔG^{o}, explain why the solubility of $Cr(OH)_3$ changes with temperature.

Explanation:

Using $\Delta G^{\theta} = \Delta H^{\theta} - T\Delta S^{\theta}$, when the temperature increases, $-T\Delta S^{\theta}$ becomes more negative. As a result, ΔG^{θ} becomes less endergonic, hence the solubility is increased.

CHAPTER 8

REDOX CHEMISTRY AND ELECTROCHEMICAL CELLS

1. (a) An aqueous solution of hydrogen peroxide, H_2O_2, decomposes in the presence of a catalyst according to the equation:

$$2H_2O_2(aq) \rightarrow 2H_2O(l) + O_2(g).$$

(i) Calculate the number of moles of H_2O_2 required to produce 10 dm^3 of oxygen gas measured at room temperature and pressure.

Explanation:

Amount of O_2 gas in 10 dm^3 = 10/24.0 = 0.417 mol.
Amount of H_2O_2 needed = 2 × 0.417 = 0.833 mol.

Do you know?

— When H_2O_2 decomposes into H_2O and O_2, it is in fact a redox reaction in disguise. It is called a *disproportionation* reaction as the same species undergo both oxidation and reduction at the same time.

— During oxidation, the *reducing agent* increases its *oxidation state* by losing electrons: $H_2O_2 \rightarrow O_2 + 2H^+ + 2e^-$.
During reduction, the *oxidizing agent* decreases its *oxidation number* by gaining electrons: $H_2O_2 + 2H^+ + 2e^- \rightarrow 2H_2O$.
The total amount of electrons released during oxidation is "consumed" during reduction. No electron is left floating around!

— How to balance a redox equation?

(*Continued*)

(Continued)

1. Assign oxidation states to determine which reactants undergo reduction and oxidation.
2. Construct two half-equations that show the specific species that are being reduced or oxidized to the corresponding products.
3. Balance each half-equation by following these simple rules:
   - balance the element that undergoes oxidation or reduction;
   - balance O atoms by adding the same number of H_2O molecules to the other side of the equation;
   - balance H atoms by adding H^+ ions to the other side of the equation; and
   - lastly, balance charges by adding electrons.
4. Repeat Step 3 for the other half-equation.
5. Ensure the number of electrons for each half-equation is the same by scaling one or both of these equations by appropriate multiples, i.e., number of e^- lost = −number of e^- gained.
6. Add the two half-equations to obtain the overall balanced equation.
7. If a basic medium is used, an OH^- ion is then added to both sides of the balanced equation to get rid of the H^+ ion in the equation.

Q Why do we need to learn how to balance a redox equation? Aren't all the half-equations given in the data booklet?

A: Well, not all are given! Thus, you need to know the basic steps of how to do it, just in case.

Q Why is "number of electrons lost = −number of electrons gained"?

A: This is because "number of electrons lost during exidation + number of electrons gained during reduction = 0." Hence, no extra electron is left floating around during a redox reaction.

Q What kind of catalyst can be used to catalyze the decomposition of H_2O_2?

A: H_2O_2 can be catalyzed by $MnO_2(s)$, Fe^{2+}, or Fe^{3+} to form H_2O and O_2. If we start off by using Fe^{2+}, then the reaction would occur as follow:

$$Fe^{2+} + H_2O_2 + 2H^+ \rightarrow Fe^{3+} + 2H_2O \qquad \text{Step 1;}$$
$$Fe^{3+} + H_2O_2 \rightarrow Fe^{2+} + O_2 + 2H^+ \qquad \text{Step 2.}$$
$$\text{Overall:} \quad 2H_2O_2 \rightarrow 2H_2O + O_2.$$

If the reaction starts off with Fe^{3+}, then Step 2 would come first, followed by Step 1. Take note that an acidic medium is necessary for the above catalytic decomposition to take place.

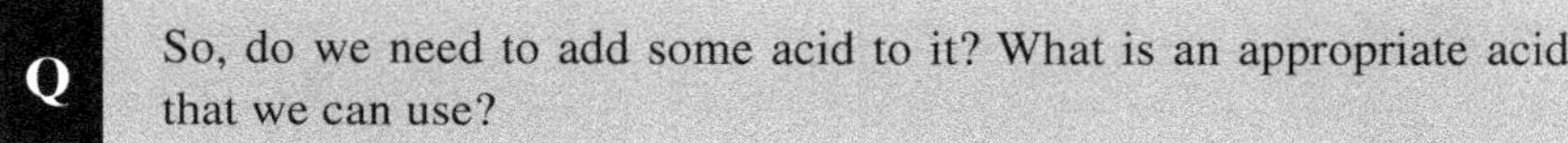

Q So, do we need to add some acid to it? What is an appropriate acid that we can use?

A: Yes, normally we use $H_2SO_4(aq)$ to acidify the medium for a redox reaction. This is because the SO_4^{2-} of H_2SO_4 is unlikely to undergo further redox reaction. We cannot use HCl or HNO_3 because the Cl^- can be easily oxidized to form Cl_2 by an oxidizing agent such as $KMnO_4$, while HNO_3 itself is an oxidizing agent.

Q But from Chapter 7, we learned that Fe^{3+} ion can undergo appreciable hydrolysis because of its high charge density, so without adding any acid, would the Fe^{3+} still be able to catalyze the decomposition?

A: Of course, the H_3O^+ that is generated from the hydrolysis of $[Fe(H_2O)_6]^{3+}$ is sufficient enough to provide the acidic medium needed:

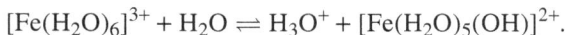

$$[Fe(H_2O)_6]^{3+} + H_2O \rightleftharpoons H_3O^+ + [Fe(H_2O)_5(OH)]^{2+}.$$

(ii) The number of moles of H_2O_2 calculated in *(a)(i)* is present in 1 dm^3 of H_2O_2 solution. Calculate the volume of this solution required to make 250 cm^3 of a 0.200 mol dm^{-3} solution, by dilution with water.

Explanation:

In 250 cm^3 of a 0.200 mol dm^{-3} H_2O_2 solution, amount of $H_2O_2 = 0.250 \times 0.2 = 0.050$ mol.

Hence, volume of 0.833 mol dm^{-3} H_2O_2 solution needed $= \frac{0.050}{0.833} = 60.0\ cm^3$.

Do you know?

— In a dilution process, the number of moles of the compound is conserved!

(b) Calculate the mass of potassium manganate (VII), $KMnO_4$, required to make 200 cm^3 of solution having a concentration of 0.040 mol dm^{-3}.

Explanation:

Molar mass of $KMnO_4 = 39.1 + 54.9 + 4 \times 16 = 158.0$ g mol^{-1}.

Amount of $KMnO_4$ in 200 cm^3 of solution having a concentration of 0.040 mol $dm^{-3} = 0.200 \times 0.04 = 0.008$ mol.

Hence, mass of $KMnO_4$ needed $= 0.008 \times 158.0 = 1.264$ g.

(c) When 20.0 cm^3 of the 0.200 mol dm^{-3} solution of H_2O_2 is acidified with sulfuric(VI) acid and titrated against a 0.040 mol dm^{-3} solution of potassium manganate (VII), $KMnO_4$, 40.0 cm^3 of the latter is required for a complete reaction.

(i) Calculate the number of moles of $KMnO_4$ in 40.0 cm^3 of the 0.040 mol dm^{-3} solution.

Explanation:

Amount of $KMnO_4$ in 40.0 cm^3 of 0.040 mol dm^{-3} solution $= 0.04 \times 0.04 = 0.0016$ mol.

(ii) Calculate the number of moles of H_2O_2 in 20.0 cm^3 of the 0.200 mol dm^{-3} solution.

Explanation:

Amount of H_2O_2 in 20.0 cm^3 of 0.200 mol dm^{-3} solution $= 0.02 \times 0.2 =$ 0.004 mol.

(iii) Hence, deduce the number of moles of H_2O_2 which reacts with 1 mole of $KMnO_4$.

Explanation:

Amount of H_2O_2 which reacts with 1 mole of $KMnO_4 = \frac{0.004}{0.0016} = 2.5$ mol.

(iv) Give a balanced equation for the reaction taking place in the titration.

Explanation:

Reaction ratio of $H_2O_2:MnO_4^- = 2.5:1 = 5:2$.

Hence, $$5H_2O_2 + 2MnO_4^- + 6H^+ \rightarrow 2Mn^{2+} + 5O_2 + 8H_2O.$$

Q What does a molar ratio of $H_2O_2:MnO_4^- = 5:2$ mean?

A: The ratio $H_2O_2:MnO_4^-$ = 5:2 means that $\frac{H_2O_2}{MnO_4^-} = \frac{5}{2}$ implying $5H_2O_2$: $2MnO_4^- = \frac{5H_2O_2}{2MnO_4^-} = \frac{1}{1}$, i.e., 5 moles of H_2O_2 would react with 2 moles of MnO_4^-. A lot of students tend to be confused with this concept, so please be careful about it!

Do you know?

— Under an acidic medium, MnO_4^- would be reduced to form Mn^{2+}! If a neutral or alkaline medium is used, the MnO_4^- would be reduced to form the insoluble, brown MnO_2:

$$MnO_4^- + 4H^+ + 3e^- \rightleftharpoons MnO_2 + 2H_2O, \qquad E^\theta = +1.67 \text{ V}.$$

— In fact, H_2O_2 is a much stronger oxidizing agent than MnO_4^-, but it cannot oxidize MnO_4^- further because the manganese atom is already in

(*Continued*)

(*Continued*)

its maximum oxidation state of +7. It cannot be further "pushed" to be above +7. So, the MnO_4^- oxidized the H_2O_2 instead because H_2O_2 can also function as a reducing agent. The relative strength of the oxidizing power of both H_2O_2 and MnO_4^- can be deduced from the standard reduction potential values:

$$MnO_4^- + 8H^+ + 5e^- \rightleftharpoons Mn^{2+} + 4H_2O, \quad E^\ominus = +1.52\text{ V}.$$
$$H_2O_2 + 2H^+ + 2e^- \rightleftharpoons 2H_2O, \quad E^\ominus = +1.77\text{ V}.$$

The more positive the $E^\ominus$ value, the more likely the species would undergo reduction, hence acting as an oxidizing agent.

(v) Explain why dilute hydrochloric acid is not used for acidification.

Explanation:

If HCl is used for acidification, the Cl^- can be oxidized by MnO_4^- to form Cl_2. Hence, the amount of MnO_4^- used would be more than what is needed to react with the H_2O_2.

(vi) Why is potassium manganate(VII) usually placed in the burette, despite the difficulties it presents in reading the burette?

Explanation:

When the colored $KMnO_4$ is added from the burette into the conical flask, it reacts with the reducing agent and converts to the near colorless Mn^{2+}. Hence, the end point of titration can be detected by one drop of excess $KMnO_4$ that turns the solution to pale pink. If we let $KMnO_4$ be in the conical flask, then it is quite difficult to titrate the solution from purple to pale pink and then to colorless. This is because the color transition from pale pink to colorless is more difficult to detect than from colorless to pale pink.

(d) The solution at the end of this reaction contains potassium sulfate (VI) and manganese (II) sulfate (VI) only.

(i) Write formulae for the cations present in this aqueous solution.

Explanation:

The formulae are K^+ and Mn^{2+}.

(ii) Treatment of the solution with dilute sodium hydroxide gives a precipitate which does not dissolve in excess sodium hydroxide solution. Identify the precipitate by name or formula.

Explanation:

The precipitate is manganese(II) hydroxide or $Mn(OH)_2$.

Do you know?

— The white $Mn(OH)_2$ is easily oxidized by air to form the brown manganese(III) hydroxide, $Mn(OH)_3$.

Q Why isn't the $Mn(OH)_2$ dissolved in excess NaOH?

A: Well, unlike other cations such as Zn^{2+}, Al^{3+}, Cr^{3+}, or Pb^{2+}, the Mn^{2+} ion does not form a soluble complex with excess OH^- ions.

2. (a) Explain the meaning of the term *oxidation* in terms of

(i) electron transfer, and

Explanation:

Oxidation refers to the process of losing electrons.

(ii) change in oxidation number (oxidation state).

Explanation:

As a result of losing electrons, oxidation results in an increase in oxidation number.

Do you know?

— Oxidation number indicates the degree of oxidation an element in a substance has. This degree of oxidation is the charge that the atom would have if the substance is broken down into individual atomic ions.

— A positive oxidation state indicates that the atom is likely to lose electrons while a negative one indicates the gaining of electrons.

— The more electronegative an atom is, the more likely it would have a negative oxidation number, i.e., gaining electrons.

— Oxidation number is not always equal to the formal charge of the species, but sometimes it may be, for example, the oxidation number of Na^+ is +1 while Cl^- is −1. But there is no formal charge for the Cl atom in HCl although it has an oxidation number of −1.

— To know the oxidation number of an atom more accurately, it would be best to know the molecular structure of the species. For example, the oxidation number that is calculated for the S atom in $S_2O_3^{2-}$ is on the average, $2x + 3(-2) = -2 \Rightarrow x = +2$. But the actual oxidation state for each of the S atom is 0 and +4, respectively:

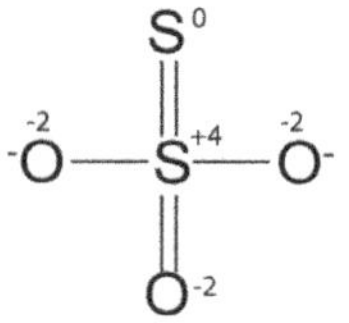

(b) What is the oxidation number of nitrogen in hydroxylamine, NH_2OH?

Explanation:

Let the oxidation number of the N atom be x.

$$x + 2 - 2 + 1 = 0 \Rightarrow x = -1.$$

Do you know?

— To calculate the oxidation number of an atom, all you need to do is just assume that the other atoms form either a cation or an anion, based on their electronegativity value. Then, sum up all the charges and equate it to the formal charge of the species, if any. Use the guideline that the electronegativity value increases across a period and decreases down the group to help you to decide which is more electronegative. The more electronegative an atom is, the more likely it would form an anion.

(c) (i) Write down the half-equations for the oxidation of iron(II) to iron(III) ions and the reduction of manganate(VII) to manganese(II) ions under acidic conditions.

Explanation:

Oxidation half-equation: $Fe^{2+} \rightarrow Fe^{3+} + e^-$.

Reduction half-equation: $MnO_4^- + 8H^+ + 5e^- \rightarrow Mn^{2+} + 4H_2O$.

Do you know?

— From the Data Booklet, we can obtain the two reduction equations:

$$MnO_4^- + 8H^+ + 5e^- \rightleftharpoons Mn^{2+} + 4H_2O, \qquad E^\ominus = +1.52\ V.$$
$$Fe^{3+} + e^- \rightleftharpoons Fe^{2+}, \qquad E^\ominus = +0.77\ V.$$

Since the equation is being presented in the reduction form, the species on the left of the equation is the oxidizing agent, as it undergoes reduction. The species on the right is the reducing agent as the backward reaction is an oxidation! From the two reduction equations, we can conclude that MnO_4^- is a stronger oxidizing agent than Fe^{3+} since the standard reduction potential for MnO_4^- is more positive than that for Fe^{3+}.

— To formulate an overall balance redox equation, all one needs is to convert the double-arrowed equation into a single-arrowed one, for the reduction half-equation:

$$MnO_4^- + 8H^+ + 5e^- \rightarrow Mn^{2+} + 4H_2O.$$

As for the oxidation half-equation, the whole reduction equation has to be reversed:

$$Fe^{2+} \rightarrow Fe^{3+} + e^-.$$

Since a balanced redox reaction means that the number of electrons released during oxidation is completely taken in during reduction, we need to balance the electron flow.

Q Why must we convert the double arrow into a single arrow when formulating the half-equations?

A: This is because in a reduction half-equation, reduction has already occurred while in an oxidation half-equation, oxidation has already happened! The reduction half-equation that we get from the Data Booklet has not "confirmed" whether it is going to undergo oxidation or reduction. Both the forward and backward reactions are likely to occur and the system is considered to be in dynamic equilibrium. But this is not the case in a redox reaction.

Q How is the standard reduction potential measured? Why is it necessary to define such a system of data?

A: (1) It is easy to understand that every species (atom, molecule, or ion) has its own unique affinity or potential to undergo reduction. Or, from an opposite perspective, the potential to undergo oxidation. So, someone decided to just look it from the reduction potential's perspective.

(2) How to measure this reduction potential? Well, you need a standard of reference. So, someone just decided to use the Standard Hydrogen Electrode or S.H.E. for short, as the reference standard:

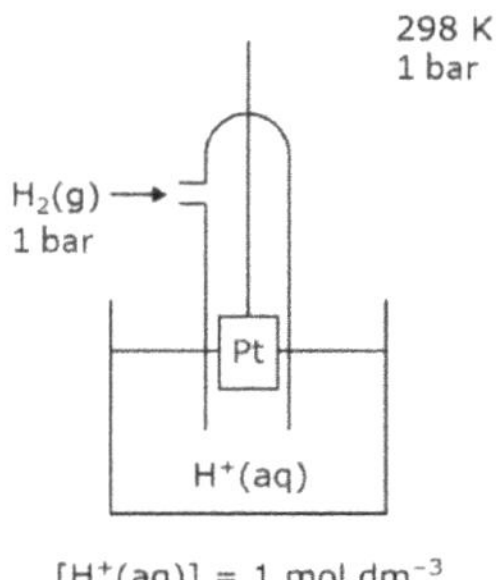

The S.H.E., which comprises the $H^+|H_2$ half cell, revolves around the equilibrium between H^+ and H_2:

$$2H^+(aq) + 2e^- \rightleftharpoons H_2(g), \quad E^\theta = 0.00 \text{ V (arbitrarily assigned value).}$$

Why is there an equilibrium being established? If you add HCl(aq) to a piece of zinc metal, what would happen to the H^+? It would react with Zn to give hydrogen gas. So, there is only one option for the H^+! How about adding H_2 gas to Cl_2, what would happen to the H_2 gas? It would convert to HCl. So, again, there is only one option for the H_2 gas. But now, if you have a mixture of H^+ and H_2 in the same system, then this system would then be able to undergo either reduction or oxidation, all depending on which other system this particular S.H.E. is being connected to. Thus, with this reference electrode, we can define the standard reduction potential as:

"E^θ of a standard half-cell is defined as the electromotive force (e.m.f.) of that half-cell relative to the standard hydrogen half-cell, under standard conditions. The more positive the E^θ value, the more likely the half-cell would undergo reduction as compared to S.H.E."

(3) Now that we have a reference standard, how are we going to physically measure the standard reduction potential with reference to the S.H.E.?

Depending on the nature of the substance, we have a total of three possible cell set-ups:

— Metal and its ion:

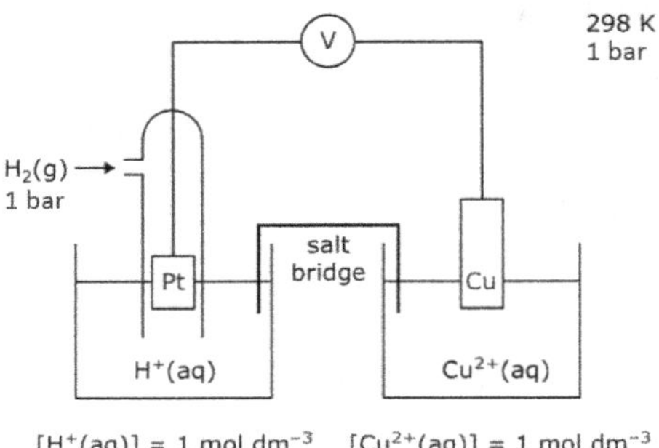

This set-up is used to measure the standard reduction potential of a metal in which the metal itself can also function as the metal electrode.

— Non-metal and its ion:

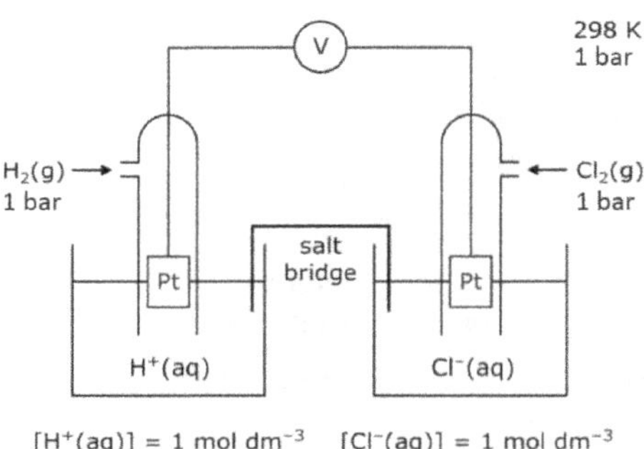

This set-up is used to measure the standard reduction potential of a non-metal, where a platinum electrode acts as an interface for the oxidation or reduction process to occur.

— Two ions of the same elements:

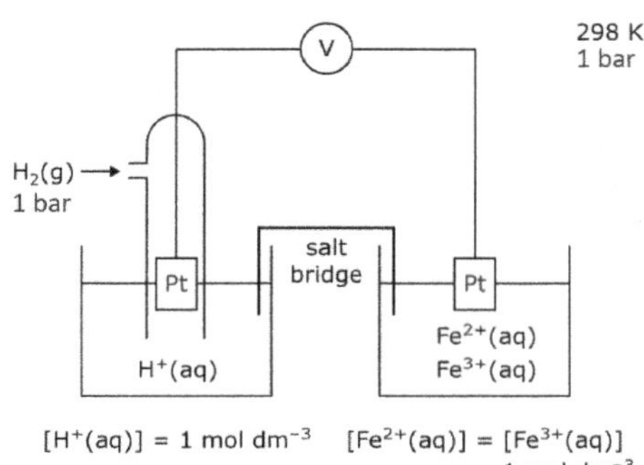

This set-up is used to measure the standard reduction potential of two ions of the same element, where a platinum electrode acts as an interface for the oxidation or reduction process to occur.

(4) All these measurements are done under standard conditions:
- — temperature of 298K;
- — pressure of gases at 1.00 bar (e.g., partial pressure of H_2 is 1.00 bar for the $H^+|H_2$ half-cell); and
- — concentration of ions at 1.00 mol dm^{-3} (e.g., $[H^+]$ is 1.0 mol dm^{-3} for the $H^+|H_2$ half-cell).

Why do we need to do all these definitions? This is because temperature, pressure, and concentration all affect our measurement. Take for instance that the E_{cell} value can be altered by changing the E values for the cathode and anode.

Example

$$Cu^{2+}(aq) + 2e^- \rightleftharpoons Cu(s),\ E^\theta = +0.34\ V.$$

The E^θ value can be made more positive by increasing $[Cu^{2+}(aq)]$ beyond 1 mol dm^{-3}.

As $[Cu^{2+}(aq)]$ increases, according to Le Chatelier's Principle, the forward reaction would be more favorable. The measured E value becomes more positive.

Conversely, if something is added to the half-cell to remove $Cu^{2+}(aq)$, such as the addition of OH^- to form the insoluble $Cu(OH)_2(s)$, the measured E value would be less positive than +0.34 V.

(5) After you have obtained the system of standard reduction potential values, how can you fully make use of it?
- — Oxidizing power

  The more positive the E^θ value, the more likely it is for a species that is on the left of the reduction equation to undergo reduction ⇒ it is more likely to be an oxidizing agent as compared to the H^+ of the S.H.E. This conclusion is obtained because when a half-cell is hooked up to the S.H.E., it was found that if this half-cell undergoes reduction while the S.H.E. undergo oxidation, then the measured potential value is positive. Similarly, the more negative the E^θ value, the more likely it is for the reduced species on the right of the reduction equation to undergo oxidation as compared to the H_2 of the S.H.E. ⇒ it is more likely to be a reducing agent.
- — Reactivity of metal

  The more negative the E^θ value of a metal cation, the more unlikely it is for the cation to be reduced ⇒ the metal element would be more likely to undergo oxidation instead.

— Calculating E^{θ}_{cell} to predict spontaneity of a redox reaction
The overall E^{θ}_{cell} of a cell set up can be calculated via

$$E^{\theta}_{cell} = E^{\theta}_{Red} - E^{\theta}_{Ox},$$

where $E^{\theta}_{Red} = E^{\theta}$value of substance reduced at the cathode and
$E^{\theta}_{Ox} = E^{\theta}$value of substance oxidized at the anode.

So, given two E^{θ} values, how do you decide which is the E^{θ}_{Red}? The more positive the E^{θ}value, the more likely the half-cell would be the cathode, hence it is E^{θ}_{Red}.

Hence, if $E^{\theta}_{cell} > 0 \Rightarrow$ redox reaction is thermodynamically spontaneous under standard conditions.

If $E^{\theta}_{cell} < 0 \Rightarrow$ redox reaction is thermodynamically non-spontaneous under standard conditions.

If $E^{\theta}_{cell} = 0 \Rightarrow$ there is no current flow and the cell has reached an equilibrium state or the cell is flat!

Q Why is there a salt bridge needed in the cell set-up?

A: The salt bridge is to maintain electrical neutrality in each of the half-cells. It consists of a small tube containing concentrated KNO_3 solution. Imagine a half-cell in which the cation in the solution undergoes reduction, the solution is going to become negatively charged because of the counter-anion that is building up. Now, the K^+ ion from the salt bridge can diffuse into this half-cell to maintain electrical neutrality. Similarly, for a half cell that undergoes oxidation, the cation is going to build up. The NO_3^- ions from the salt bridge can then diffuse into the half-cell to neutralize the positive charge has been built up.

Q Why when the $E^{\theta}_{cell} > 0$, is the redox reaction thermodynamically spontaneous under standard conditions?

A: The calculated E^{θ}_{cell} value is related to the Gibbs Free Energy through $\Delta G^{\theta} = -n\text{F}E^{\theta}_{cell}$, where n is the number of moles of electrons that flow during the redox reaction in the balanced redox reaction and F is Faraday's Constant. If $E^{\theta}_{cell} > 0$, then $\Delta G^{\theta} < 0$, and the redox reaction is thermodynamically spontaneous.

Q Can there be a redox reaction that is predicted to occur but does not happen? If so, why?

A: If a redox reaction is predicted to occur but does not happen, it may be because it is thermodynamically feasible (i.e., $E^{\theta}{}_{cell}/E^{\theta} > 0$) but kinetically non-feasible because of (i) high activation energy or (ii) reaction conditions are non-standard.

(ii) Deduce the ionic equation for the reaction between iron(II) ions and manganate(VII) ions under acidic conditions.

Explanation:

The balance redox equation: $MnO_4^- + 8H^+ + 5Fe^{2+} \rightarrow Mn^{2+} + Fe^{3+} + 4H_2O$.

(d) The following experiment was used to determine the equation for the reaction between hydroxylamine, NH_2OH, and iron(III) ions. 0.740 g of hydroxylamine was dissolved in water and made up to 50.0 cm^3. The solution was reacted with an excess solution of an acidified iron(III) salt. When the reaction was completed, the iron(II) produced required 44.8 cm^3 of 0.200 mol dm^{-3} potassium manganate(VII) solution to oxidize the iron(II) back to iron(III).

(i) Calculate the amount of hydroxylamine used in the reaction.

Explanation:

Molar mass of hydroxylamine = 14 + 2 +16 + 1 = 33 g mol^{-1}.
Amount of hydroxylamine used in the reaction = 0.74/33 = 0.0224 mol.

(ii) Calculate the amount of iron(II) formed in the reaction.

Explanation:

Amount of MnO_4^- used = $0.0448 \times 0.200 = 8.96 \times 10^{-3}$ mol.

$$MnO_4^- + 8H^+ + 5Fe^{2+} \rightarrow Mn^{2+} + Fe^{3+} + 4H_2O$$

Amount of $Fe^{2+} = 5 \times 8.96 \times 10^{-3} = 4.48 \times 10^{-2}$ mol.

(iii) Determine the molar ratio of iron(III) to hydroxylamine reacting together.

Explanation:

Amount of Fe^{3+} = Amount of Fe^{2+} = 4.48×10^{-2} mol.
Molar ratio of $Fe^{3+}:NH_2OH = 4.48 \times 10^{-2}:0.224 \approx 2:1$.

Do you know?

The molar ratio of $Fe^{3+}:NH_2OH = 2:1$ means that $2Fe^{3+}:NH_2OH = 1:1$ or $\frac{2Fe^{3+}}{NH_2OH} = \frac{1}{1}$.

(iv) Using both parts *(b)* and *(d)(iii)*, deduce the oxidation number of nitrogen in the product.

Explanation:

Oxidation state of N in NH_2OH is $x + 2 - 2 + 1 = 0 \Rightarrow x = -1$.

Let the final oxidation state of N in the product be y.

Amount of e^- released during oxidation = $-$Amount of e^- consumed during reduction.

$\Rightarrow 1 \times (y - (-1)) = -2 \times (2 - 3)$

$\Rightarrow y = +1$.

Hence, the oxidation number of nitrogen in the product is +1.

Do you know?

— To determine the initial or final oxidation number of a reactant or product, respectively, all you need to do is to just make use of the basic concept "Amount of e^- released during oxidation = $-$Amount of e^- consumed during reduction."

(Continued)

(*Continued*)

— To calculate the change in oxidation state during oxidation, simply use the "final oxidation state minus initial oxidation state," and taking the sign of the oxidation state into consideration during the calculation. While to calculate the change in oxidation state during reduction, also use the "final oxidation state minus the inital oxidation state." But the latter calculation has to multiply by a negative sign. Why? This is because in a redox reaction, the "number of electrons lost in oxidation + number of electrons gained in reduction = 0, hence number of electron lost in oxidation = – number of electrons gained in reduction.

(v) Which of the following possible nitrogen containing compounds, NO, N_2O, N_2O_4, N_2, and NH_3, is the most likely product of the reaction?

Explanation:

The oxidation number of N in: NO is –2; N_2O is +1; N_2O_4 is +4; N_2 is 0; and NH_3 is –3.

Hence, N_2O is the most likely product.

(vi) Write the equation for the reaction between hydroxylamine and iron(III) ions.

Explanation:

$$2NH_2OH + 4Fe^{3+} \rightarrow N_2O + 4Fe^{2+} + H_2O + 4H^+$$

Do you know?

— The half-equation for the oxidation of NH_2OH to N_2O is not given in the Data Booklet. But you can derive it on your own as follows:

(1) Identify the reacting species and its product: $NH_2OH \rightarrow N_2O$
(2) Balance the reacting species: $2NH_2OH \rightarrow N_2O + H_2O$
(3) Balance O atom by adding H_2O: $2NH_2OH \rightarrow N_2O + H_2O$
(4) Balance H atom by adding H^+: $2NH_2OH \rightarrow N_2O + H_2O + 4H^+$
(5) Balance the charge using e^-: $2NH_2OH \rightarrow N_2O + H_2O + 4H^+ + 4e^-$

3. A possibility for the future is to use electric cars that generate their own electricity. Instead of burning methanol in an internal combustion engine, methanol could be used to power a *fuel cell* in the car. The fuel cell generates an electric current which drives an electric motor. The fuel and oxygen are fed continuously to two electrodes immersed in an acidic electrolyte solution. The electrodes are made of platinum dispersed onto a porous carbon support.

   (a) Write half-equations for the reactions which take place at the anode and the cathode of the fuel cell, and combine these to give an overall equation for the cell reaction.

Explanation:

Cathode reaction: $O_2 + 4H^+ + 4e^- \rightarrow 2H_2O$.
Anode reaction: $CH_3OH + H_2O \rightarrow CO_2 + 6H^+ + 6e^-$.

Do you know?

— Since oxidation occurs at the anode where electrons are being released, the electrons will flow from the anode to the cathode. Hence, the anode is negatively charged while the cathode is positively charged.

— If O_2 gas is used as the oxidizing agent in such a cell as above, then the cathode reaction is either this half–equation $O_2 + 4H^+ + 4e^- \rightarrow 2H_2O$ or this half–equation $O_2 + 2H_2O + 4e^- \rightarrow 4OH^-$.

— An acidic medium is required for the above reaction as H^+ ions are needed!

Q Can we write the cathode reaction as $O_2 + 2H_2O + 4e^- \rightarrow 4OH^-$ here?

A: No, you can't. This is because the medium used is acidic in nature. If the medium used is neutral or alkaline, then you can use $O_2 + 2H_2O + 4e^- \rightarrow 4OH^-$.

Q Which medium is preferred, acidic or alkaline?

A: The acidic medium is preferred because:

$$O_2 + 4H^+ + 4e^- \rightleftharpoons 2H_2O, \qquad E^\theta = +1.23\ V.$$
$$O_2 + 2H_2O + 4e^- \rightleftharpoons 4OH^-, \qquad E^\theta = +0.40\ V.$$

You would get a higher $E^\theta_{cell} = E^\theta_{Red} - E^\theta_{Ox}$ when you use an acidic medium because the E^θ_{Red} used is more positive!

Q But is there any disadvantage if we use an acidic medium?

A: Well, the acid used may corrode the metal parts of the set-up.

Q Why are the electrodes made of platinum dispersed onto a porous carbon support?

A: There is no point using a piece of solid platinum as the electrode, as platinum is expensive. The dispersion of the platinum metal over the porous carbon support increases the surface area for the active platinum to work. In addition, the carbon support is also conducting in nature. Hence, there is no problem for electrons to flow through the electrode.

Q How does a fuel cell set-up looks like?

A: You can draw the set-up as follows if you use NaOH as the electrolyte and hydrogen as the fuel:

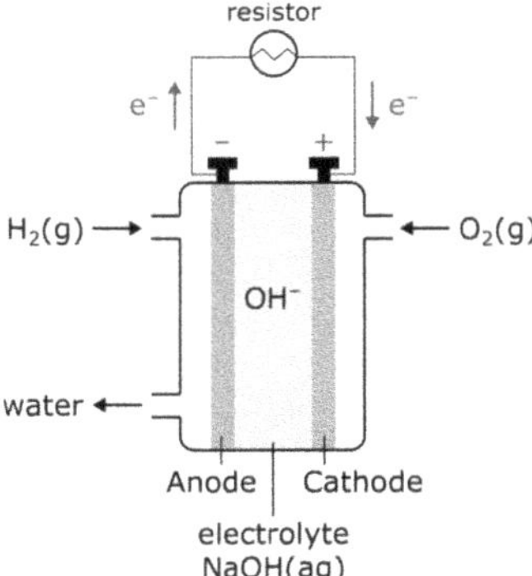

In fact, you can use fuels such as hydrazine (NH_2NH_2), ethanol, or sugar solution. It will still work.

Q How do you derive the oxidation half-equation for the CH_3OH fuel?

A: Well, you can do it as follows:

(1) Identify the reacting species and its product: $CH_3OH \rightarrow CO_2$
(2) Balance the reacting species: $CH_3OH \rightarrow CO_2$
(3) Balance O atom by adding H_2O: $CH_3OH + H_2O \rightarrow CO_2$
(4) Balance H atom by adding H^+: $CH_3OH + H_2O \rightarrow CO_2 + 6H^+$
(5) Balance the charge using e^-: $CH_3OH + H_2O \rightarrow CO_2 + 6H^+ + 6e^-$

Q How about the oxidation half-equation for NH_2NH_2 if we use an alkaline medium?

A: (1) Identify the reacting species and its product: $NH_2NH_2 \rightarrow N_2$
(2) Balance the reacting species: $NH_2NH_2 \rightarrow N_2$
(3) Balance H atom by adding H^+: $NH_2NH_2 \rightarrow N_2 + 4H^+$
(4) Balance the charge using e^-: $NH_2NH_2 \rightarrow N_2 + 4H^+ + 4e^-$
(5) Use OH^- to balance the H^+: $NH_2NH_2 + 4OH^- \rightarrow N_2 + 4H_2O + 4e^-$

(b) Explain where the energy which powers the electric motor comes from.

Explanation:

The energy which powers the electric motor comes from the stored chemical energy of the reactants. Thus, the fuel cell converts chemical energy to electrical energy, which is then further converted to mechanical energy.

Do you know?

— A battery, or a voltaic cell or a galvanic cell, stores chemical energy which is released as electrical energy during a redox reaction.

(Continued)

(*Continued*)

— The larger the amount of electrons that is released during oxidation would mean that the more electrical energy we would get from the reaction. So, for example, sugar gives us more energy during oxidation as compared to methanol:

$$CH_3OH + H_2O \rightarrow CO_2 + 6H^+ + 6e^-$$
$$C_6H_{12}O_6 + 6H_2O \rightarrow 6CO_2 + 24H^+ + 24e^-.$$

(c) Explain why the fuel cell is much more environmentally sound than a conventional internal combustion engine.

Explanation:

Conventional internal combustion engines release other gases such as NO and NO_2, which pollute the environment. The fuel cell that we are discussing here only releases CO_2 and water.

Do you know?

— If H_2 is used as the fuel, the product is H_2O, which is environmentally non-polluting.
— If hydrazine, NH_2NH_2, is used, we have N_2 and H_2O as the products.

(d) One refinement of the fuel cell design is to replace the acidic electrolyte solution with a film of solid H^+ ion-conducting electrolyte. Explain why this would be an improvement.

Explanation:

Liquid acid tends to be more corrosive than a film of solid H^+ ion-conducting electrolyte. In addition, it is easy to handle a fuel cell that utilizes less liquid substances.

4. Tin cans made of tin-plated iron are used to preserve food. Tin has the advantage in that it corrodes much less readily than iron and it forms a protective layer preventing the iron from rusting. When the coating is scratched, however, the iron rusts faster when it is in contact with tin. Fortunately, neither Fe^{2+} nor Sn^{2+} ions are toxic.

   (a) State *two* substances which are necessary for iron to rust and from which iron is protected by the tin layer.

Explanation:

Iron needs oxygen and water in order to rust. Thus, tin excludes the iron from coming in contact with these two substances.

Do you know?

— When iron rusts, the Fe(s) is converted to Fe^{2+} and Fe^{3+}:

$$Fe \rightarrow Fe^{2+} + 2e^-$$
$$Fe^{2+} \rightarrow Fe^{+} + e^-.$$

The electrons are then taken in by O_2 and H_2O in the following:

$$O_2 + 2H_2O + 4e^- \rightarrow 4OH^-.$$

This explains why O_2 and H_2O are needed.

Q Why is tin less readily oxidized than iron?

A: From the standard reduction potential values,

$$Fe^{2+} + 2e^- \rightleftharpoons Fe, \quad E^\theta = -0.44\ V.$$
$$Sn^{2+} + 2e^- \rightleftharpoons Sn, \quad E^\theta = +0.15\ V.$$

Since the reduction of Sn^{2+} is more readily than the reduction of Fe^{2+}, this would mean that Sn^{2+} is a stronger oxidizing agent. Conversely, Sn would be a weaker reducing agent than Fe. The logic is very simple! If an oxidizing agent "likes" to be reduced, then the reduced form must NOT like to be oxidized back! In another perspective, when $Sn \rightleftharpoons Sn^{2+} + 2e^-$ its $E^\theta = -0.15$ V, while for $Fe \rightleftharpoons Fe^{2+} + 2e^-$, its $E^\theta = +0.44$ V Since the standard oxidation potential for Fe is more positive than that of Sn, Fe is likely to be oxidized.

(b) Write a balanced equation for the reaction where the presence of tin ions encourages the iron to corrode.

Explanation:

Reduction half–equation: $Sn^{2+} + 2e^- \rightarrow Sn$.
Oxidation half–equation: $Fe \rightarrow Fe^{2+} + 2e^-$.
Overall reaction: $Fe + Sn^{2+} \rightarrow Fe^{2+} + Sn$.

Do you know?

— If you calculate the E^θ_{cell} value for the above redox reaction, you would get:

$$E^\theta_{cell} = E^\theta_{Red} - E^\theta_{Ox} = +0.15 - (-0.44) = +0.59 \text{ V.}$$

The reaction is thermodynamically spontaneous under standard conditions.

Q How come $E^\theta = +0.44$ wasn't used for the E^θ_{Ox} value since it is an oxidation process?

A: Well, you DON'T reverse the sign for E^θ_{Ox} when you use $E^\theta_{cell} = E^\theta_{Red} - E^\theta_{Ox}$. Just substitute in as what you have gotten from the Data Booklet. The subscripts 'Red' and 'Ox' ONLY indicate reduction (or cathode) and oxidation (or anode), respectively. BUT if you use $E^\theta_{cell} = E^\theta_{Red} + E^\theta_{Ox}$ instead, then for the value that you have gotten from the Data Booklet for the anode reaction, you need to reverse it. For example:

If you reverse the sign for $E^\theta = -0.44$ V to become $E^\theta = +0.44$ V and substitute into $E^\theta_{cell} = E^\theta_{Red} + E^\theta_{Ox} = +0.15 + (+0.44) = +0.59$ V, you still get the same answer.

Q If we are asked to draw the cell set-up for the above reaction, how would it look like?

A: Well, it looks like the following:

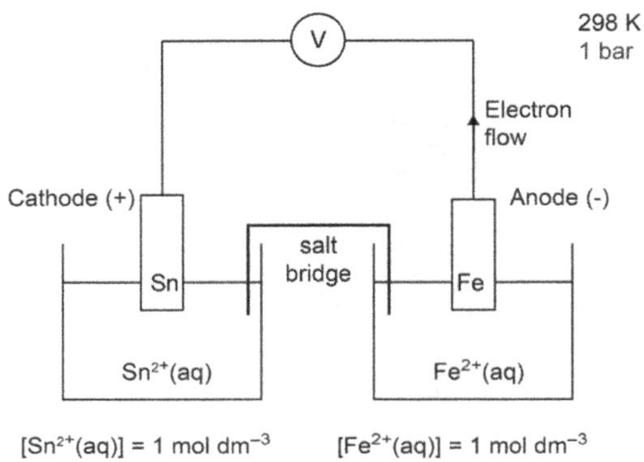

The e.m.f. of the set-up is obtained simply by connecting the positive terminal of the voltmeter to the positive terminal (cathode) of the cell set-up. Similarly, connect the negative terminal to the anode. The e.m.f. value is the first reading shown by the voltmeter the moment the circuit is completed, corresponding to the e.m.f. value under standard conditions. After taking the first reading, you would notice that the e.m.f. would decrease. This is because the concentrations of the species are no longer at 1 mol dm^{-3}.

(c) Write a balanced equation for the corrosion of iron.

Explanation:

Oxidation half-equation: $Fe \rightarrow Fe^{2+} + 2e^{-}$.

Reduction half-equation: $O_2 + 2H_2O + 4e^{-} \rightarrow 4OH^{-}$.

Overall reaction: $2Fe + O_2 + 2H_2O \rightarrow 2Fe^{2+} + 4OH^{-}$.

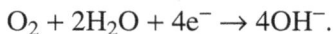

Q Why did you not use this equation, $O_2 + 4H^+ + 4e^- \rightleftharpoons 2H_2O$ $E^\theta = +1.23$ V, instead?

A: We are assuming that it is a non-acidic medium. But if the contents in the can are acidic in nature, then we should use this equation instead.

Q So, this would mean that iron would corrode faster in an acidic medium?

A: Yes, indeed. The O_2 will more readily undergo reduction in an acidic medium as shown by the more positive E^θ value of +1.23 V. In addition, if there are other electrolytes such as NaCl present, the corrosion rate is accelerated as the electrolyte helps to increase the electrical conductivity of the medium. That is why iron corrodes faster near the seaside where the wind is moist with NaCl.

(d) Rust is often given the formula $Fe_2O_3.xH_2O$, where x has a variable non-integral value. Calculate the value of x for a sample of rust which loses 22% mass (as steam) when heated to constant mass.

Explanation:

In 100g of $Fe_2O_3.xH_2O$, there are 22g of H_2O.
Amount of H_2O in 22g = 22/18 = 1.22 mol.
Molar mass of $Fe_2O_3.xH_2O$ = 2(55.8) + 3(16) + x(18) = (159.6 + 18x) g mol^{-1}.
In 100g of $Fe_2O_3.xH_2O$, amount of $Fe_2O_3.xH_2O = \frac{100}{(159.6+18x)}$ mol.
Hence, amount of H_2O in 100 g $= x \cdot \frac{100}{(159.6+18x)} = 1.22$

$$\Rightarrow 100x = 1.22 \times (159.6 + 18x)$$

$$\Rightarrow x = 2.50$$

(e) An underground iron pipe is less likely to corrode if bonded at intervals to magnesium stakes. Give a reason for this. Explain why aluminium would be a poor substitute for the magnesium.

Explanation:

$$Mg^{2+} + 2e^- \rightleftharpoons Mg, \quad E^\theta = -2.38\text{ V.}$$

$$Fe^{2+} + 2e^- \rightleftharpoons Fe, \quad E^\theta = -0.44\text{ V.}$$

$$Al^{3+} + 3e^- \rightleftharpoons Al, \quad E^\theta = -1.66\text{ V.}$$

The standard reduction potential value for the reduction of Mg^{2+} is more negative than that of Fe^{2+}, which means that the reduction of Mg^{2+} is less feasible than that of the Fe^{2+}. Hence, this shows that Mg metal is more likely to undergo oxidation than Fe metal. If Al metal is used instead, although Al metal is also more likely to be oxidized than the Fe metal, but there will be layer of Al_2O_3 coated on the surface, which is impervious to both water and oxygen. Hence, the Al metal underneath the oxide layer cannot further be oxidized. Therefore, the Al metal cannot protect the Fe metal from being corroded.

5. This question concerns the lead-acid battery. The following data will be required.

	$E^{\ominus}/V$
$PbO_2(s) + 4H^+(aq) + SO_4^{2-}(aq) + 2e^- \rightleftharpoons PbSO_4(s) + 2H_2O(l)$	+1.69
$PbSO_4(s) + 2e^- \rightleftharpoons Pb(s) + SO_4^{2-}(aq)$	−0.36

(a) The lead-acid battery is one form of storage cell. What substance is used for:

(i) the negative pole;

Explanation:

For a battery, the negative pole refers to the anode. According to the given standard reduction potential values, the following are the cathode and anode reactions:

Cathode: $PbO_2(s) + 4H^+(aq) + SO_4^{2-}(aq) + 2e^- \rightarrow PbSO_4(s) + 2H_2O(l)$.
Anode: $Pb(s) + SO_4^{2-}(aq) \rightarrow PbSO_4(s) + 2e^-$.

The negative pole is the Pb metal electrode.

Do you know?

— In an electrochemical cell, which includes both a voltaic and an electrolytic cell, a metal or an electrically conducting substance needs to be used as the electrode so that electrons can flow through.

(ii) the positive pole; and

Explanation:

There is no metal in the half-equation but there is PbO_2:

$$PbO_2(s) + 4H^+(aq) + SO_4^{2-}(aq) + 2e^- \rightleftharpoons PbSO_4(s) + 2H_2O(l).$$

So, a logical electrode for the positive pole would be a piece of Pb metal, coated with a thin layer of PbO_2.

Do you know?

— We cannot use the lead metal oxide as the electrode as it is non-conducting in nature. But if there is another metal oxide that can conduct electricity, then it would be possible to employ it as the electrode.

(iii) the electrolyte.

Explanation:

The appropriate electrolyte to use is $H_2SO_4(aq)$.

(b) Give the equation for the overall cell reaction during discharge.

Explanation:

Cathode: $PbO_2(s) + 4H^+(aq) + SO_4^{2-}(aq) + 2e^- \rightarrow PbSO_4(s) + 2H_2O(l)$.

Anode: $Pb(s) + SO_4^{2-}(aq) \rightarrow PbSO_4(s) + 2e^-$.

Overall reaction: $PbO_2(s) + Pb(s) + 4H^+(aq) + 2SO_4^{2-}(aq) \rightarrow 2PbSO_4(s) + 2H_2O(l)$.

(c) Calculate the e.m.f. of the cell.

Explanation:

The e.m.f. of the cell = $E^{\theta}_{cell} = E^{\theta}_{Red} - E^{\theta}_{Ox} = +1.69 - (-0.36) = +2.05$ V.

(d) A storage cell, as used in the lead-acid battery, is a simple cell in which the reactions are reversible, i.e., once the chemicals have been used up they can be re-formed. Write an equation for the chemical reaction which occurs on charging.

Explanation:

The equation during charging would be reversed during discharging:

$$2PbSO_4(s) + 2H_2O(l) \rightarrow PbO_2(s) + Pb(s) + 4H^+(aq) + 2SO_4^{2-}(aq)$$

(e) Give one disadvantage of such batteries used in cars.

Explanation:

When the battery can no longer be used, the lead inside the battery is an environmental hazard.

Do you know?

— When we charge a battery, we are actually converting electrical energy back to chemical energy. The process is very similar to that of an electrolytic cell.

— During electrolysis, not all species in the system can undergo oxidation or reduction. In fact, there is selective discharge of species, which depends on the E^{θ} value. The more positive the E^{θ} value,

(Continued)

(Continued)

⇒ the more likely the oxidized state will undergo reduction at the cathode but

⇒ the less likely the reduced form will undergo oxidation at the anode.

— The sign of the cathode and anode is opposite to those in the voltaic cell. In an electrolytic cell, the cathode is connected to the negative terminal of the power source. Hence, species in the system can only go to this negative terminal to "take in" electrons, i.e., undergo reduction. The anode on the other hand, is connected to the positive terminal of the power source, so species can only go to the electrode to "deposit" electrons, i.e., undergo oxidation!

— During electrolysis, the amount of material that is discharged at each of the electrodes depends solely on the amount of current that passes through the system. Thus, using the following relationship, we can perform quantitative calculations for an electrolytic reaction:

$Q = I.t$ or $Q = nF$ where Q = quantity of charge in coulombs
I = current in amperes
t = time in seconds
F = Faraday's Constant (96500 C mol^{-1})
n = number of moles of electrons

— There are three electrolytic cells that we need to be familiar with:

(1) Electroplating or purification of copper

Anode: $Cu(s) \rightarrow Cu^{2+}(aq) + 2e^-$.

Cathode: $Cu^{2+}(aq) + 2e^- \rightarrow Cu(s)$.

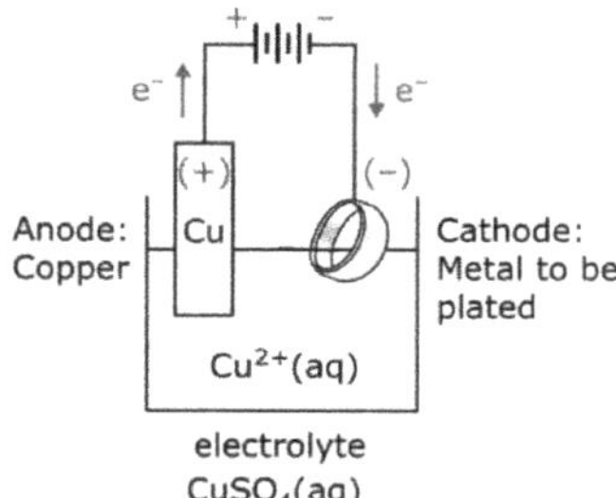

	Anode	Cathode
Electroplating:	Copper	Metal to be plated
Purification:	Impure copper	Pure copper

(Continued)

(Continued)

(2) Anodization of aluminium

Anode (Al metal): First, $2H_2O(l) \rightarrow O_2(g) + 4H^+(aq) + 4e^-$.

Then, $4Al(s) + 3O_2(g) \rightarrow 2Al_2O_3(s)$.

Cathode (Graphite): $2H^+(aq) + 2e^- \rightarrow H_2(g)$.

The electrolyte used is dilute $H_2SO_4(aq)$.

(3) Electrolysis of brine (concentrated sodium chloride solution)

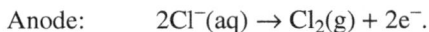

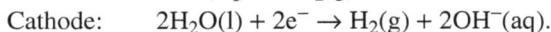

Anode: $2Cl^-(aq) \rightarrow Cl_2(g) + 2e^-$.

Cathode: $2H_2O(l) + 2e^- \rightarrow H_2(g) + 2OH^-(aq)$.

Important for producing bleach:

$2NaOH(aq) + Cl_2(aq) \rightarrow NaClO(aq) + NaCl(aq) + H_2O(l)$.

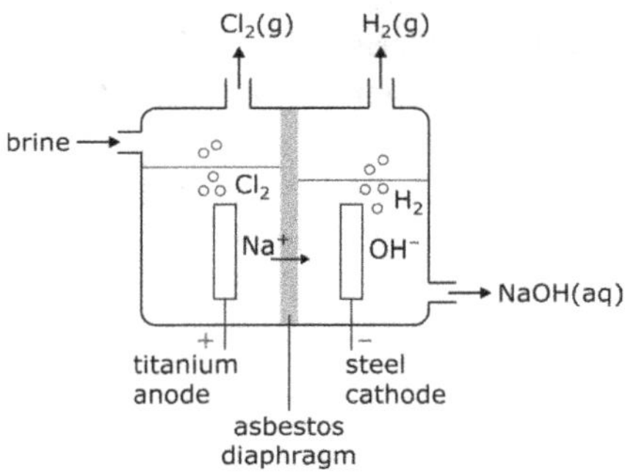

Q How can we remember the sign of the electrode in the voltaic cell and the electrolytic cell without any confusion?

A: All you need to do is remember that the anode is an oxidation electrode where electrons are being released, while the cathode is where reduction takes place and electrons are being consumed.

Q Why didn't the SO_4^{2-}, H^+, or H_2O molecules undergo reduction at the cathode during the purification of copper?

A: Here are the E^θ values for:

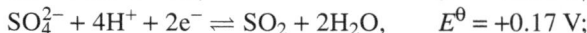

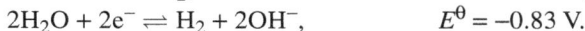

$Cu^{2+} + 2e^- \rightleftharpoons Cu$, $E^\theta = +0.34$ V;

$SO_4^{2-} + 4H^+ + 2e^- \rightleftharpoons SO_2 + 2H_2O$, $E^\theta = +0.17$ V;

$2H^+ + 2e^- \rightleftharpoons H_2$, $E^\theta = +0.00$ V;

$2H_2O + 2e^- \rightleftharpoons H_2 + 2OH^-$, $E^\theta = -0.83$ V.

According to the E^θ values, the reduction of Cu^{2+} is thermodynamically more spontaneous than the rest.

Why didn't the OH^- and H_2O undergo discharge at the anode during the purification of copper?

A: Here are the E^θ values for:

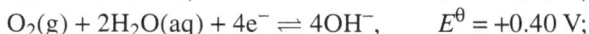

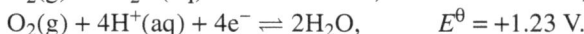

$Cu^{2+} + 2e^- \rightleftharpoons Cu$, $E^\theta = +0.34$ V;
$O_2(g) + 2H_2O(aq) + 4e^- \rightleftharpoons 4OH^-$, $E^\theta = +0.40$ V;
$O_2(g) + 4H^+(aq) + 4e^- \rightleftharpoons 2H_2O$, $E^\theta = +1.23$ V.

According to the E^θ values, the oxidation of Cu metal is thermodynamically more spontaneous than the rest.

During anodization of aluminium, can we use plain water instead of acidic solution?

A: No! The electrical conductivity of plain water is not as good as acidic solution which contains ions as mobile charge carries.

Why didn't the OH^- and the Al undergo oxidation first at the anode during the anodization of aluminium?

A: Here are the E^θ values for:

$Al^{3+} + 3e^- \rightleftharpoons Al$, $E^\theta = -1.66$ V;
$O_2(g) + 2H_2O(aq) + 4e^- \rightleftharpoons 4OH^-$, $E^\theta = +0.40$ V;
$O_2(g) + 4H^+(aq) + 4e^- \rightleftharpoons 2H_2O$, $E^\theta = +1.23$ V.

According to the E^θ values, Al metal is the most likely "candidate" to undergo oxidation at the anode. But the moment a layer of Al_2O_3 is formed, the Al metal cannot be further oxidized. So, what happens is that H_2O becomes the next likely candidate to be oxidized as it is present in a larger amount as compared to the meager amount of OH^- ions present in the acidic solution. So, remember that if we use an acidic solution, the H_2 gas actually comes from H^+ and the O_2 gas actually comes from H_2O and not OH^-.

Since Al_2O_3 is amphoteric in nature, why didn't it react with the acidic aqueous H_2SO_4?

A: This is because the anode is positive charged, it would repel the H^+ ion away. So, how can the Al_2O_3 reacts with H^+?

Q What is the purpose of the asbestos diaphragm in the electrolysis of brine?

A: The asbestos diaphragm allows the brine electrolyte to diffuse from the anode chamber to the cathode chamber without any back-mixing.

Q Why didn't the OH^- and the H_2O undergo oxidation at the anode during the electrolysis of brine?

A: Here are the E^θ values for:

$Cl_2 + 2e^- \rightleftharpoons 2Cl^-$, $E^\theta = +1.36$ V;
$O_2(g) + 2H_2O(aq) + 4e^- \rightleftharpoons 4OH^-$, $E^\theta = +0.40$ V;
$O_2(g) + 4H^+(aq) + 4e^- \rightleftharpoons 2H_2O$, $E^\theta = +1.23$ V.

According to the E^θ values, Cl^- is the least likely to undergo oxidation. But because the brine solution is concentrated in nature, Cl^- is more likely to be oxidized. Comparatively, although OH^- is more likely to undergo oxidation than H_2O, its concentration is far too low compared to that of the H_2O molecules that are present.

Q Why didn't the H^+ and the Na^+ undergo reduction at the cathode during the electrolysis of brine?

A: Here are the E^θ values for:

$2H^+ + 2e^- \rightleftharpoons H_2$, $E^\theta = +0.00$ V;
$Na^+ + e^- \rightleftharpoons Na$, $E^\theta = -2.71$ V;
$2H_2O + 2e^- \rightleftharpoons H_2 + 2OH^-$, $E^\theta = -0.83$ V.

According to the E^θ values, Na^+ is unlikely to undergo reduction as the reduction is thermodynamically non-spontaneous. As for H^+, the concentration is too low for its reduction to be significant.

Q In the electrolysis of brine, why is it that both electrodes that are used cannot be either titanium or steel?

A: Titanium is more expensive than steel. The anode has to be titanium because it is more resistant to the corrosive Cl_2 gas than steel. As for the cathode, a steel electrode is good enough as it would not be reactive to the cathodic products.

PART II

INORGANIC CHEMISTRY

CHAPTER 9

THE PERIODIC TABLE: CHEMICAL PERIODICITY

1. (a) Elements in the *p* block of the periodic table show great variation in physical and chemical properties. Explain this variation, with reference to the properties of aluminium, silicon, phosphorous, and chlorine.

Explanation:

From Al to Si to P to Cl, the electronegativity value increases. This means that the elements from Al to Cl would be less likely to lose electrons but instead, they would be more likely to gain or share electrons. The increase in electronegativity is due to an increase in the effective nuclear charge (ENC) acting on the valence electrons.

The Al atom has three valence electrons which are not very strongly attracted, hence these three electrons can delocalize to form metallic bonding, accounting for the high melting point of Al metal and its electrical conductivity. In addition, it is also energetically possible for Al to form an ionic compound simply by losing these three valence electrons. But due to the high charge density of the Al^{3+} ion, some of the aluminum compound is actually covalent in nature, for example $AlCl_3$.

The Si atom has four valence electrons which are relatively more strongly attracted than those in Al due to a greater ENC. As a result, it is energetically challenging for Si to lose all four electrons to form a +4 charge cation. Neither is it energetically feasible for Si to gain another four electrons to achieve the octet configuration, as the inter-electronic repulsion would be too strong for the nucleus to hold on these four extra electrons. As a result, in elemental silicon, each Si atom is tetrahedrally covalent bonded to four other Si atoms, accounting for the high melting point needed to break the strong Si–Si bond. Si atom is also likely to form covalent compounds with other elements.

With an even higher ENC and possessing five valence electrons, it is natural for a P atom to gain another three electrons to achieve the octet configuration through the formation of a −3 charge anion or through covalent bond formation with other elements. And with the presence of vacant low-lying *d* orbitals, it is energetically feasible for a P atom to form a maximum of five bonds. Hence, a P atom can have a maximum oxidation state of +5. In the elemental form, each P atom forms three covalent bonds with three other P atoms, resulting in a simple molecular structure consisting of P_4 molecules. Thus, elemental phosphorus has a low melting point because of the weak intermolecular forces.

As for the Cl atom, possessing seven valence electrons and being the most electronegative atom amongst all, it is common for the Cl atom to gain an extra electron to achieve the octet configuration through the formation of a −1 charge anion or through covalent bond formation. Hence, elemental chlorine is a diatomic molecule with a very low melting point and is also a strong oxidizing agent. Like P atom, the Cl atom can also expand beyond the octet configuration to form a compound with a maximum oxidation state of +7, utilizing its vacant low-lying *d* orbitals.

Q Why does the ENC increase across the period?

A: Across the period, the nuclear charge increases but the number of inner core electrons remains the same. Hence, the shielding effect provided by these inner core electrons is relatively constant. This causes an increase in the net electrostatic attractive force acting on the valence electrons.

Q Why are some of the aluminum compounds ionic while others are covalent in nature?

A: Whether the aluminum compound is ionic or not depends very much on the counter-anion. What do we mean by that? If an Al^{3+} is near a Cl^- ion, due to the high charge density $\left(\propto \frac{q_+}{r_+}\right)$ of the Al^{3+} ion and the high polarizability (ability to be polarized) of the Cl^- ion; the electron cloud of the Cl^- ion would be distorted to an extent that when you observe $AlCl_3$, you would find that it is a simple molecular compound and not an ionic one! But now, if the

Al^{3+} ion is near a F^- ion, due to the low polarizability of the F^- ion, there would be very minimal distortion of electron density of the F^- ion. Hence, you would observe AlF_3 as an ionic compound.

Q Why does NCl_5 not form?

A: The N atom comes from Period 2 with only the *s* and *p* subshells in its valence shell. Hence, there can only be a maximum of eight electrons in the valence shell. If the N atom "wants" to form five bonds, it needs to promote some of the valence electrons to the next energy level, which is the higher energy $n = 3$ principal quantum shell. From the "investment-return theory," this is energetically not feasible. But for the formation of PCl_5, because there are empty *d* orbitals which are close in energy level to the *p* orbitals, from the "investment-return theory," it is energetically viable to form PCl_5.

Q Why does P^{3-} form but Si^{4-} does not form?

A: The nucleus of P atom has one more proton than that of Si. Both P^{3-} and Si^{4-} are isoelectronic, therefore the interelectronic repulsion is the same. The extra proton in P^{3-} is better able to "contain" the inter-electronic in P^{3-} than Si^{4-}. Hence, P^{3-} is energetically more stable than Si^{4-}. You can use the same argument to understand why P^{3-} is less stable than S^{2-}. In a nutshell, whether a particular anionic species is formed or not depends on the balance between the inter-electronic repulsion and the attractive force acting on those electrons by the nucleus. If there is a net attractive force, then the species would be likely to form.

Do you know?

— An increase in the ENC across Period 3, increases electronegativity (or decreases electropositivity), decreases atomic radius, and increases ionization energy. So, how are the trends of physical and chemical properties like?

(1) Nature of the elements and the types of compound that they form
Across the period, the elements vary from metal to metalloid to non-metal.
As a result, electropositivity (ability to lose electrons) decreases but electronegativity (ability to gain electrons) increases.

(Continued)

(Continued)

Thus, the elements vary from forming ionic oxides/chlorides on the left to covalent oxides/chlorides on the right.
The maximum oxidation state corresponds to the number of valence electrons the atom has!

(2) Atomic radius and ionic radius

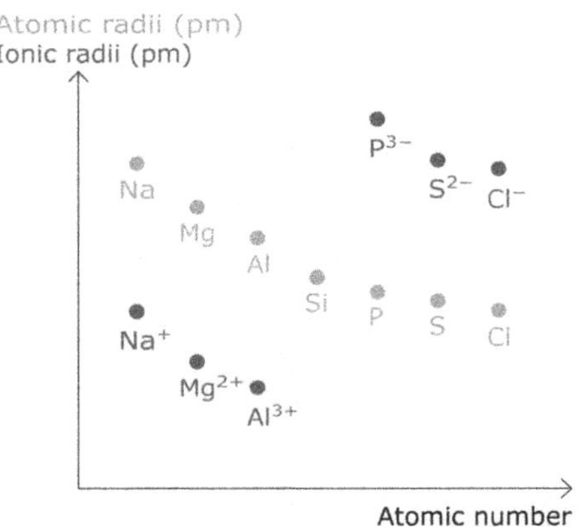

- Atomic radius decreases across the period due to an increase in the (ENC). The ENC increases because nuclear charge increases, while the shielding effect by inner core electrons is relatively constant across the period.
- Cationic radius decreases from Na^+ to Mg^{2+} to Al^{3+}. Since these species are isoelectronic (same number of electrons), the inter-electronic repulsion is the same but there is an increase in nuclear charge. Hence, the net attractive force acting on the valence electrons increases.
- Anionic radius decreases from P^{3-} to S^{2-} to Cl^-. Since these species are isoelectronic, the inter-electronic repulsion is the same but there is an increase in nuclear charge. Hence, the net attractive force acting on the valence electrons increases.
- Period 3 cations are smaller than anions from the same period; anion has an additional principal quantum shell of electrons than the cation, hence accounting for greater inter-electronic repulsion.

(3) First ionization energy

- First I.E. generally increases across the period due to an increase in ENC.

(Continued)

(Continued)

- Factors affecting the first I.E. across the period: (i) ENC, (ii) inter-electronic repulsion, and (iii) *p* electron has higher energy than *s* electron.

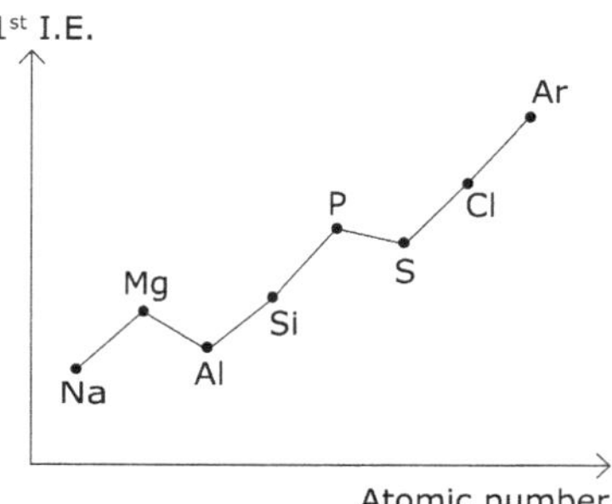

(4) Melting point
- Melting point increases from Na to Mg to Al due to an increase in metallic bond strength. This arises because of an increase in the number of valence electrons for delocalization and a decrease in atomic size.
- Si has high melting point due to the strong Si–Si covalent bond.
- Melting point decreases from S_8 to P_4 to Cl_2 to Ar due to a decrease in the number of electrons the molecule/atom has, hence a decrease in the strength of instantaneous dipole–induced dipole interactions.

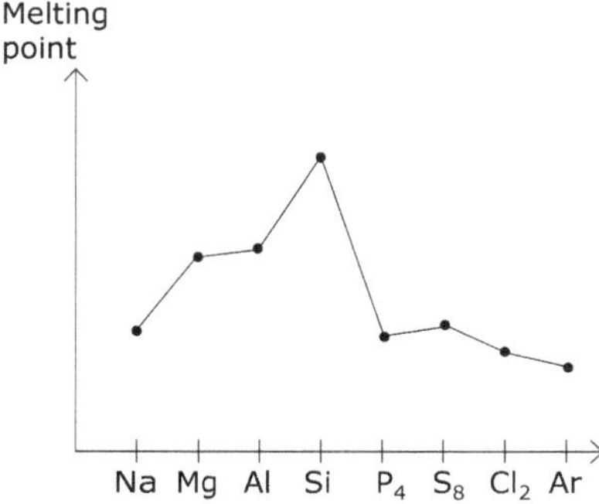

(5) Electrical conductivity
- Electrical conductivity increases from Na to Mg to Al due to an increase in the number of valence electrons for delocalization.

(Continued)

(*Continued*)

- Si is a semi-conductor.
- P, S, Cl, and Ar are non-conducting due to the absence of mobile charge carriers.

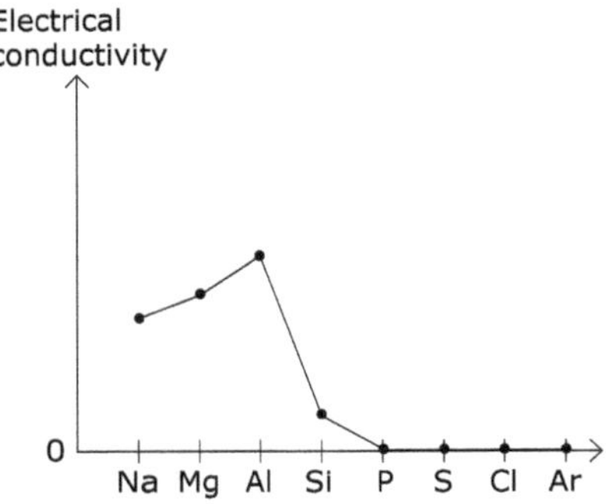

— The greater the difference between the electronegativities of two atoms, the more likely these two atoms would form an ionic compound. But if the charge density of the cation is too large and the polarizability of the anion is significant, a covalent compound will result. Hence, a metal and non-metal would not necessary from an ionic compound!

— The strength of a metallic bond depends on two factors: (1) number of delocalized electrons — this factor accounts for why melting point increases from Na to Mg to Al; and (2) size of the atom — this factor accounts for why the melting point decreases down Group 1 and Group 2.

(b) For any two of these elements, explain how their large-scale uses are determined by their physical and chemical properties.

Explanation:

Al metal has good thermal and electrical conductivities. Due to its light weight, aluminum and its alloys are used in the making of aircrafts, power lines, satellite dishes, trains, boats, cars, and building materials. As the layer of Al_2O_3 is impervious to a lot of substances like water, it is commonly use in the canning industry.

Chlorine is a gas that has relatively good solubility in water. Due to its strong oxidizing power, it is commonly used as disinfectant in the treatment of water. In addition, it is also widely used in the paper industry to bleach pulps.

Do you know?

— Silicon is widely used in the semi-conductor industry due to its low electrical conductivity. SiO_2 is commonly found in sand and it is important for glass–making.

— Phosphorous is widely used in the synthesis of fertilizers and matches. Phosphorous in the form of phosphate, PO_4^{3-}, is an important component in the making of DNA, RNA, ATP, etc.

(c) Normal electric wiring consists of copper wires surrounded by polyvinyl chloride (PVC). In one type of electric wiring used in fire alarm systems, a copper wire is surrounded by solid magnesium oxide acting as an insulator, and then encased in a copper tube covered with PVC.

(i) What type of bonding is present in magnesium oxide? Explain how it can act as an insulator.

Explanation:

MgO is an ionic compound with strong ionic bond between the Mg^{2+}cation and O^{2-} anion. This strong ionic bond accounts for the high melting point of MgO. In addition, as the ions are rigidly held in the giant ionic lattice, they cannot function as charge carriers. This accounts for its insulating property.

Do you know?

— Generally, a metal and a non-metal react to form an ionic compound. But this is not always true, for example, $AlCl_3$ and PbO_2.

— The strength of an ionic bond is dependent on: (1) the charges of the cation and anion; and (2) the ionic radii of the cation and anion. All these factors are lumped together in the lattice energy formula, L.E. $\propto \frac{q_+ \cdot q_-}{(r_+ + r_-)}$, which measures the strength of an ionic bond. Take note that L.E. indicates the strength of ionic bond but it does not explain the strength of the ionic bond! A lot of students simply say that "the ionic bond is stronger because it has a more exothermic lattice energy". This is incorrect! It should be "the more exothermic L.E. indicates that the ionic bond is stronger due to greater charges or smaller ionic radii".

(ii) Suggest why magnesium oxide is preferred to PVC as an insulator in fire alarm systems.

Explanation:

PVC would melt during a fire, hence exposing the live copper wires, which are dangerous. As for MgO, due to its high melting point, it can still protect the copper wires during a fire.

Do you know?

— Due to the high melting point of MgO, together with Al_2O_3, they are common materials used in the making of furnace linings due to their high refractory property.

2. Aluminum chloride occurs in both the anhydrous state and in the hydrated state. The structure of the anhydrous state may be perceived as having the formula $AlCl_3$. When water is added to solid anhydrous aluminum chloride, white acidic fumes are seen.

   (a) Name the white acidic fumes.

Explanation:

The white acidic fume is HCl gas.

Do you know?

— $AlCl_3$ is a non-polar, simple molecular compound with weak instantaneous dipole–induced dipole interaction.

— As the Al atom of $AlCl_3$ has an unused *p* orbital, this Al atom is highly electron-deficient. Hence, the $AlCl_3$ can form the Al_2Cl_6 dimer:

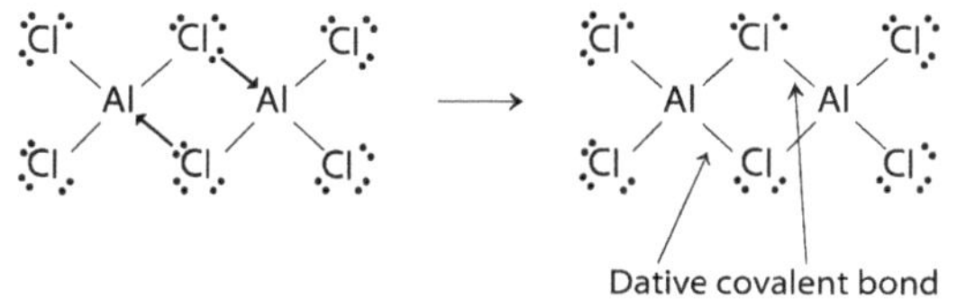

Dative covalent bond

As a result, the Al atom changes from a trigonal planar shape, which is a sp^2 hybridized state in $AlCl_3$, to a tetrahedral shape, which is a sp^3 hybridized state in Al_2Cl_6. The bond angle also changes from 120° to 109.5°.

— Both Al_2Cl_6 and $AlCl_3$ are non-polar, but since Al_2Cl_6 has more electrons than $AlCl_3$, the instantaneous dipole–induced dipole interaction is stronger in Al_2Cl_6. Hence, when we compare the meeting point of aluminium chloride with other compounds, make sure that we know whether we are using $A1C1_3$ or $A1_2C1_6$. The intermolecular forces for $A1_2C1_6$ is stronger than that of $A1C1_3$.

(b) Write an equation for the reaction occurring.

Explanation:

$$AlCl_3(s) + 3H_2O(l) \rightarrow Al(OH)_3(s) + 3HCl(g).$$

Q Can we write the equation as, $2AlCl_3(s) + 3H_2O(l) \rightarrow Al_2O_3(s) + 6HCl(g)$?

A: How do you think the HCl is formed? Where does it come from? You can imagine the lone pair of electrons on the O atom of the H_2O being attracted to the highly electron-deficient Al atom of the $AlCl_3$. Do not forget that there is an empty *p* orbital on this Al atom. So, as the electron-deficient Al atom attracts the lone pair of electrons on the O atom, the intramolecular O–H bond weakens. This causes a HCl molecule to be "expelled," since the H atom of the H_2O molecule is also in close proximity to the Cl atom. Now, after the H_2O molecule has lost a H^+ ion to become OH^-, in order to form the O^{2-} in Al_2O_3, the OH^- ion needs to lose another H^+ ion. So, do you think it is easy to remove a H^+ ion from a OH^- ion? Not really! Therefore, it is less likely to form Al_2O_3 under this condition. But if you further heat the $Al(OH)_3$ that is formed, then it would undergo dehydration to give Al_2O_3:

$$2Al(OH)_3(s) \rightarrow Al_2O_3(s) + 3H_2O(g).$$

Q Since, a H_2O molecule is broken up to form a HCl, can we say that the above reaction is a hydrolysis reaction?

A: Yes, certainly. The reaction $AlCl_3(s) + 3H_2O(l) \rightarrow Al(OH)_3(s) + 3HCl(g)$ is in fact a hydrolysis reaction.

Do you know?

(1) Types of Period 3 chlorides

— For the Period 3 elements, the rate of forming the chlorides corresponds to the ability of the element to lose the valence electrons. The more likely the element is able to lose electrons, the more likely it would form the ionic chloride.

— The maximum oxidation number corresponds to the number of valence electrons the element has. From phosphorus onward, the elements are able to expand their octet configuration due to the "availability of empty low-lying vacant *d* orbitals." Hence, these elements are able to promote their valence electrons from the 3*p* subshell to the 3*d* subshell.

(*Continued*)

(Continued)

Element	Reaction with chlorine	Structure and bonding of the chloride	Possible oxidation state
Na	Reacts vigorously to form neutral NaCl $2Na(s) + Cl_2(g) \rightarrow 2NaCl(s)$	Ionic lattice with strong ionic bond	+1
Mg	Reacts vigorously to form weakly acidic $MgCl_2$ $Mg(s) + Cl_2(g) \rightarrow MgCl_2(s)$	Ionic lattice with strong ionic bond	+2
Al	Reacts vigorously to form weakly acidic $AlCl_3$ $Al(s) + \frac{3}{2} Cl_2(g) \rightarrow AlCl_3(s)$ $2Al(s) + 3Cl_2(g) \rightarrow Al_2Cl_6(s)$	Simple molecular compound with weak intermolecular forces between the molecules	+3
Si	Reacts slowly to form acidic $SiCl_4$ $Si(s) + 2Cl_2(g) \rightarrow SiCl_4(l)$	Simple molecular compound with weak intermolecular forces between the molecules	+4
P	Reacts slowly to form acidic PCl_3 and PCl_5 $P(s) + \frac{3}{2} Cl_2(g) \rightarrow PCl_3(l)$ $P(s) + \frac{5}{2} Cl_2(g) \rightarrow PCl_5(s)$	Simple molecular compound with weak intermolecular forces between the molecules	+3, +5

(Continued)

(Continued)

(2) Melting point of the Period 3 chlorides

— Melting point decreases from NaCl to $MgCl_2$ due to the presence of covalent character in $MgCl_2$, which weakens the ionic bond. This arises because of the higher charge and smaller cationic radius of the Mg^{2+} ion as compared to Na^+.

— Melting point decreases from Al_2Cl_6 to PCl_5 to $SiCl_4$ due to a decrease in the strength of instantaneous dipole–induced dipole interaction as the number of electrons decreases.

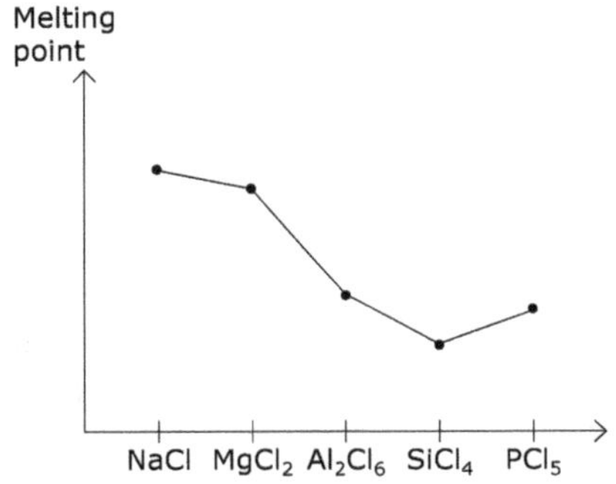

(3) Chemical properties of the Period 3 chlorides in water

— Non-metal covalent chlorides, such as $SiCl_4$, PCl_3, and PCl_5, react with H_2O to give a weak acidic solution because the electron-deficient Si or P atom is able to attract the lone pair of electrons of a H_2O molecule to form covalent bond with the vacant low-lying *d* orbitals of the central atom. The attractive force weakens the O–H bond of the H_2O molecule and causes it to break.

— NaCl dissolves in water to form aqueous NaCl where pH = 7:

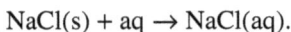

$$NaCl(s) + aq \rightarrow NaCl(aq).$$

— $MgCl_2$ dissolves in water to give a weakly acidic solution where pH ≈ 6.5. The hydrolysis occurs because of the high charge density of the Mg^{2+} ion:

$$MgCl_2(s) + 6H_2O(l) \rightarrow [Mg(H_2O)_6]^{2+}(aq) + 2Cl^-(aq)$$
$$[Mg(H_2O)_6]^{2+} + H_2O \rightleftharpoons [Mg(H_2O)_5(OH)]^+(aq) + H_3O^+.$$

(Continued)

(*Continued*)

— $AlCl_3$ dissolves in water to give a weakly acidic solution where pH ≈ 3. The extent of hydrolysis is so high that the addition of carbonate results in the evolution of CO_2:

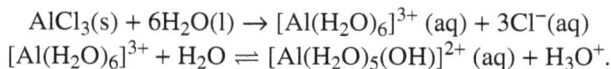

$$AlCl_3(s) + 6H_2O(l) \rightarrow [Al(H_2O)_6]^{3+}(aq) + 3Cl^-(aq)$$
$$[Al(H_2O)_6]^{3+} + H_2O \rightleftharpoons [Al(H_2O)_5(OH)]^{2+}(aq) + H_3O^+.$$

— $SiCl_4$, PCl_3, and PCl_5 hydrolyze in water to give a highly acidic solution of pH ≈2. The high acidity is a result of the formation of HCl, which is a strong acid:

$$SiCl_4(l) + 2H_2O(l) \rightarrow SiO_2(s) + 4HCl(aq);$$
$$PCl_3(s) + 3H_2O(l) \rightarrow H_3PO_3(aq) + 3HCl(aq); \text{ and}$$
$$PCl_5(s) + 4H_2O(l) \rightarrow H_3PO_4(aq) + 5HCl(aq).$$

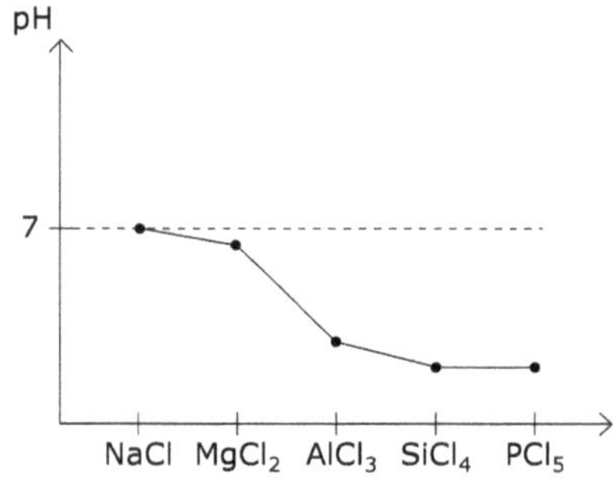

The reaction of PCl_5 is highly exothermic! The presence of cold water or a limited amount of water can limit further hydrolysis of $POCl_3$. When the mole ratio of PCl_5:H_2O = 1:1 or the water added is cold or in a limited amount, we would have:

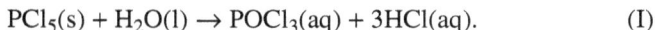

$$PCl_5(s) + H_2O(l) \rightarrow POCl_3(aq) + 3HCl(aq). \quad \text{(I)}$$

When more water is now added:

$$POCl_3(aq) + 3H_2O(l) \rightarrow H_3PO_4(aq) + 2HCl(aq). \quad \text{(II)}$$

When excess water is added AND the water used is not cold:

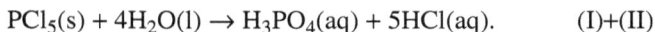

$$PCl_5(s) + 4H_2O(l) \rightarrow H_3PO_4(aq) + 5HCl(aq). \quad \text{(I)+(II)}$$

Q Why does the rate of reaction of the Period 3 elements reacting with Cl_2 decreases across the period?

A: Did you notice that the greater the difference between the electronegativity values of the Period 3 elements and chlorine, the faster the rate of reaction? So, the rate decreases because toward the end of the period, the electronegativity differences between the elements and chlorine get smaller and smaller.

Q Why would the presence of covalent character weaken the ionic bond in $MgCl_2$?

A: Covalent character causes the electron cloud to be distorted toward the cation. As a result, the net charge on the cation decreases. Similarly, the net charge on the anion would also decrease. Hence, since ionic bond strength depends on the charges of the ions, the ionic bond strength decreases.

Q Why are the central atom of these Period 3 chlorides, $SiCl_4$, PCl_3, and PCl_5, electron-deficient?

A: The central atoms of these chlorides are bonded to some highly electronegative Cl atoms, hence making them electron-deficient. Therefore, they are able to attract the lone pair of electrons from the water molecule. Plus, as they have vacant low-lying d orbitals, the oxygen atom of the H2O molecule can form covalent bond with them.

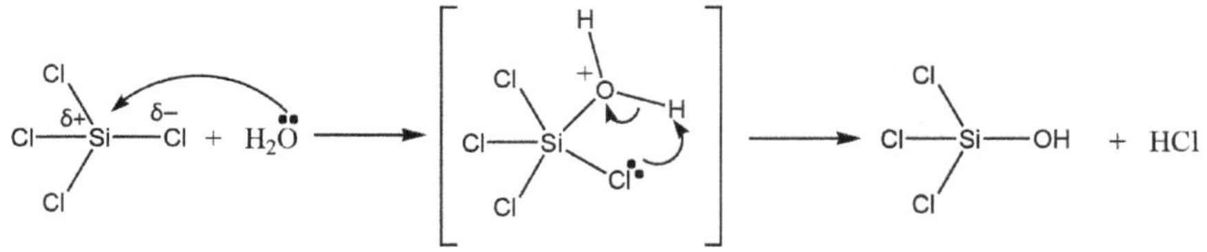

Q The carbon atom of CCl_4 is also electron-deficient like $SiCl_4$. Why doesn't CCl_4 undergo hydrolysis like $SiCl_4$?

A: Yes, the carbon atom of CCl_4 is indeed electron-deficient but unfortunately, the H_2O molecule cannot approach the carbon atom close enough to form covalent bond with it, due to the steric effect exhibit by the four bulky Cl atoms. In addition, the carbon atom does not have vacant low-lying orbitals to form bonds with the water molecule. These bring to our attention that in order to understand whether a reaction proceeds or not, it is important to look at all the possible factors that affect the reaction!

Q If $Al_2(CO_3)_3$ does not exist, does $MgCO_3$ exist?

A: Yes, $MgCO_3$ does exist. The acidity causes by the hydrolysis of the $[Mg(H_2O)_6]^{2+}$ complex is not strong enough to decompose the carbonate ion.

Q Why isn't the hydrolysis equation of $SiCl_4$ written as: $SiCl_4(l) + 4H_2O(l) \rightarrow Si(OH)_4(s) + 4HCl(aq)$? Like the one for $AlCl_3$: $AlCl_3(s) + H_2O(l) \rightarrow Al(OH)_3(s) + HCl(g)$?

A: For Si and P atoms, these elements' electronegativities are not much different from that of the O atom. As a result, imagine four –OH groups covalently bonded to a Si atom or five –OH groups bonded to a P atom. The electronegative Si or P atoms are going to be highly electron-deficient. Hence, spontaneous dehydration is going to take place. That is why for a high-oxidation-state compound containing too many O atoms, they are in the form called oxo-compound, which includes, PO_4^{3-}, SO_4^{2-}, MnO_4^-, etc.

(c) Explain with reference to the structure of $AlCl_3$ how the first step of this reaction occurs.

Explanation:

The Al atom of $AlCl_3$ has a vacant *p* orbital and it is highly electron-deficient. The lone pair of electrons on the O atom of the H_2O molecule is attracted to this highly electron-deficient Al atom. So, as the electron-deficient Al atom attracts the lone pair of electrons on the O atom, the intramolecular O–H bond weakens.

(d) When water is added to hydrated aluminum chloride, no white fumes are observed. Explain this in terms of the bonding in hydrated aluminum chloride.

Explanation:

In hydrated aluminum chloride, the aluminum is present as the $[Al(H_2O)_6]^{3+}$ complex and the compound is ionic in nature. Thus, when water is added, the compound dissolves to give the $[Al(H_2O)_6]^{3+}$ complex and Cl^- ions.

(e) Hydrated aluminum chloride dissolves in water to give an acidic solution. Explain why the solution is acidic with an appropriate equation.

Explanation:

Due to the high charge density of the Al^{3+} ion, the Al^{3+} ion has high polarizing power. The electron cloud of the water molecules are polarized to the extent that the O–H bond is weakened. This results in the release of free H_3O^+ ion:

$$[Al(H_2O)_6]^{3+} + H_2O \rightleftharpoons [Al(H_2O)_5(OH)]^{2+} + H_3O^+.$$

3. NO_2 reacts with aqueous sodium hydroxide according to the following equation:

$$2OH^- + 2NO_2 \rightarrow NO_2^- + NO_3^- + H_2O.$$

(a) What type of reaction is this? Justify your answer.

Explanation:

It is a disproportionation reaction. The oxidation state of N in NO_2 is +4; in NO_2^- is +3; and in NO_3^- is +5. Thus, NO_2 is oxidized to NO_3^- and is also reduced to NO_2^-.

Do you know?

— Both CO_2 and SO_2 also reacts with OH^- as follows:

$$2OH^- + CO_2 \rightarrow CO_3^{2-} + H_2O, \text{ and}$$
$$2OH^- + SO_2 \rightarrow SO_3^{2-} + H_2O.$$

But these two reactions are acid–base reactions! This is because both CO_2 and SO_2 are acidic gases while OH^- is a base.

(b) Deduce the ionic half-equations for this reaction.

Explanation:

Oxidation half-equation: $NO_2 + 2OH^- \rightarrow NO_3^- + H_2O + e^-$.
Reduction half-equation: $NO_2 + e^- \rightarrow NO_2^-$.

4. (a) Sodium hydroxide is manufactured by an electrolytic process using a diaphragm cell.
   (i) What is used as the electrolyte?

Explanation:

The electrolyte is concentrated sodium chloride solution.

(ii) What material is the anode and cathode made of, respectively?

Explanation:

Titanium anode and steel cathode.

Do you know?

— Titanium metal has to be used as the electrode for the anode in the electrolysis of brine as the Cl_2 gas that is evolved in the anode compartment is corrosive in nature. On the other hand, H_2 gas in the cathode compartment would not react with the steel electrode. Hence, it is okay using the cheaper steel metal.

(iii) Give an equation for the reaction occurring at each of the electrodes.

Explanation:

Anode: $2Cl^-(aq) \rightarrow Cl_2(g) + 2e^-$.
Cathode: $2H_2O(l) + 2e^- \rightarrow H_2(g) + 2OH^-(aq)$.

(iv) Give one reason why it is necessary to separate the two electrodes in to two compartments.

Explanation:

Cl_2 gas and H_2 gas react very explosively. Hence, they need to be separated during their production.

Do you know?

— The enthalpy change for the reaction:

$$H_2(g) + Cl_2(g) \rightarrow 2HCl(g)$$

$$\text{is } \Delta H = \text{BE(H–H)} + \text{BE(Cl–Cl)} - 2\ \text{BE(H–Cl)}$$
$$= 436 + 244 - 2(431) = -182\,\text{kJ mol}^{-1}.$$

(i) Write an equation for the overall cell reaction.

Explanation:

Overall cell reaction: $2H_2O(l) + 2Cl^-(aq) \rightarrow H_2(g) + 2OH^-(aq) + Cl_2(g)$.

Do you know?

— $OH^-(aq)$ is important for the production of bleaching agent, $ClO^-(aq)$:

$$2OH^-(aq) + Cl_2(g) \rightarrow ClO^-(aq) + Cl^-(aq) + H_2O(l).$$

(b) Give one large-scale industrial use for each of the following:
(i) chlorine; and

Explanation:

Chlorine is important in the production of the bleaching agent, NaClO(aq).

Do you know?

— The bleach, NaClO(aq), works by oxidizing the coloured molecule into a colorless one, while itself being reduced to the chloride ion.

(ii) hydrogen.

Explanation:

Hydrogen is important in the catalytic hydrogenation of unsaturated fats to saturated fats.

Do you know?

— Hydrogenated fats have a higher melting point than that of the unsaturated fats as the saturated molecules have less double bonds, hence resulting in a greater surface area for the formation of instantaneous dipole–induced dipole interaction due to the more linear structure of the molecule. But saturated fats are not good for the body as they can solidify in the arteries and cause clotting.
— Hydrogenated fats have a longer shelf-life as there are less double bonds. Hence, they are less susceptible to attack by other chemicals.

(a) Iron(II) oxide is a basic oxide. What type of oxide is:

(i) aluminum oxide; and

Explanation:

Aluminum oxide is an amphoteric oxide, meaning it can react with both an acid and a base:

$$Al_2O_3(s) + 6H^+(aq) \rightarrow 2Al^{3+}(aq) + 3H_2O(l)$$

$$Al_2O_3(s) + 2OH^-(aq) + 3H_2O(l) \rightarrow 2[Al(OH)_4]^-(aq).$$

Do you know?

— $Al_2O_3(s)$ can react with H^+ because the O^{2-} is highly electron-rich, hence it is basic enough to react with the electron-deficient H^+.

— $Al_2O_3(s)$ can react with OH^- because the Al^{3+} has high charge density, hence it is electron-deficient enough to react with OH^-. The charge density is so high that Al^{3+} can react with four OH^- to form the $[Al(OH)_4]^-(aq)$ complex.

— There are other amphoteric oxides such as:

ZnO: $ZnO(s) + 2H^+(aq) \rightarrow Zn^{2+}(aq) + H_2O(l)$

$ZnO(s) + 2OH^-(aq) + H_2O(l) \rightarrow [Zn(OH)_4]^{2-}(aq)$;

PbO: $PbO(s) + 2H^+(aq) \rightarrow Pb^{2+}(aq) + H_2O(l)$

$PbO(s) + 2OH^-(aq) + H_2O(l) \rightarrow [Pb(OH)_4]^{2-}(aq)$;

BeO: $BeO(s) + 2H^+(aq) \rightarrow Be^{2+}(aq) + H_2O(l)$

$BeO(s) + 2OH^-(aq) + H_2O(l) \rightarrow [Be(OH)_4]^{2-}(aq)$; and

Cr_2O_3: $Cr_2O_3(s) + 6H^+(aq) \rightarrow 2Cr^{3+}(aq) + 3H_2O(l)$

$Cr_2O_3(s) + 6OH^-(aq) + 3H_2O(l) \rightarrow 2[Cr(OH)_6]^{3-}(aq)$.

So, the common reason why the metal cation of each of the oxide above is able to react with a strong base is due its high charge density.

(ii) silicon dioxide?

Explanation:

Silicon dioxide is a giant covalent compound. It reacts with a strong base such as NaOH(aq) to give the sodium silicate:

$$SiO_2(s) + 2NaOH(aq) \rightarrow Na_2SiO_3(aq) + H_2O(l).$$

Do you know?

— Among all the common acids, SiO_2 can only be is attacked by HF:
$SiO_2(s) + 6HF(aq) \rightarrow H_2SiF_6(aq) + 2H_2O(l)$.
Thus, HF is commonly used in the semi-conductor industry to etch away SiO_2.
HF cannot be stored in glass bottle but HCl(aq), HBr(aq), and HI(aq) can be.

— The reason why SiO_2 can react with NaOH is because the Si atom that is bonded to four highly electronegative O atoms, is highly electron-deficient. Hence, it "attracts" attack from the strong base.

(d) Bauxite an ore containing hydrated aluminum oxide, iron(III) oxide, and silicon dioxide. In order to obtain a purer form of aluminum oxide, bauxite is heated with a 10% solution of sodium hydroxide in which the aluminum oxide dissolves.

(i) Write an equation for the reaction of aluminum oxide with sodium hydroxide.

Explanation:

$$Al_2O_3(s) + 2NaOH(aq) + 3H_2O(l) \rightarrow 2Na[Al(OH)_4](aq).$$

(ii) Why does iron(III) oxide not dissolve in sodium hydroxide?

Explanation:

The charge density of the Fe^{3+} is not high enough for it to form a soluble complex with sodium hydroxide.

(iii) Why does silicon dioxide not dissolve in the 10% solution sodium hydroxide?

Explanation:

The concentration of the sodium hydroxide is not high enough to cause soluble $Na_2SiO_3(aq)$ to form.

Do you know?

(1) Types of Period 3 oxides

— The rate of formation of the oxides corresponds to the ability of the element to lose its valence electrons. The more likely the element is able to lose electrons, the more likely the formation of the ionic oxide.

— The maximum oxidation number corresponds to the number of valence electrons the element has. From phosphorus onward, elements are able to expand their octet configuration due to the "availability of empty low-lying vacant *d*-orbitals." Hence, elements are able to promote their valence electrons from the 3*p* subshell to the 3*d* subshell.

Element	Reaction with oxygen	Structure and bonding of the oxide	Possible oxidation state
Na	Reacts vigorously to form basic oxide Na_2O. $2Na(s) + \frac{1}{2}O_2(g) \rightarrow Na_2O(s)$	Ionic lattice with strong ionic bond	+1

(*Continued*)

(*Continued*)

Element	Reaction with oxygen	Structure and bonding of the oxide	Possible oxidation state
Mg	Reacts vigorously to form basic oxide MgO $Mg(s) + \frac{1}{2}O_2(g) \rightarrow MgO(s)$	Ionic lattice with strong ionic bond	+2
Al	Vigorous reaction initially but the oxide layer formed soon prevents further reaction $2Al(s) + \frac{3}{2}O_2(g) \rightarrow Al_2O_3(s)$	Ionic lattice with strong ionic bond	+3
Si	Reacts slowly to form SiO_2 $Si(s) + O_2(g) \rightarrow SiO_2(s)$	Giant covalent structure with strong covalent bond	+4
P	Reacts vigorously to form P_4O_6 and P_4O_{10} depending on reaction conditions $P_4(s) + 3O_2(g) \rightarrow P_4O_6(s)$ $P_4(s) + 5O_2(g) \rightarrow P_4O_{10}(s)$	Simple molecular compound with weak intermolecular forces between the molecules	+3, +5
S	Reacts slowly to form SO_2 which oxidizes very slowly to SO_3 without a catalyst $S(s) + O_2(g) \rightarrow SO_2(g)$ $2SO_2(g) + O_2(g) \rightarrow 2SO_3(g)$	Simple molecular compound with weak intermolecular forces between the molecules	+4, +6

(2) Melting point of the Period 3 oxides

(*Continued*)

(Continued)

— Melting point increases from Na_2O to MgO due to an increase in the ionic bond strength. This arises because of the higher charge and smaller cationic radius of the Mg^{2+} ion as compared to that of Na^+.
— The lower melting point of Al_2O_3 is due to the high charge density of the Al^{3+} ion which gives rise to a high polarizing power. This leads to the presence of covalent character which weakens the ionic bond strength.
— The high melting point of SiO_2 is due to strong Si–O covalent bond.
— Melting point decreases from P_4O_{10} to SO_3 due to a decrease in the strength of instantaneous dipole–induced dipole as the number of electrons decreases.

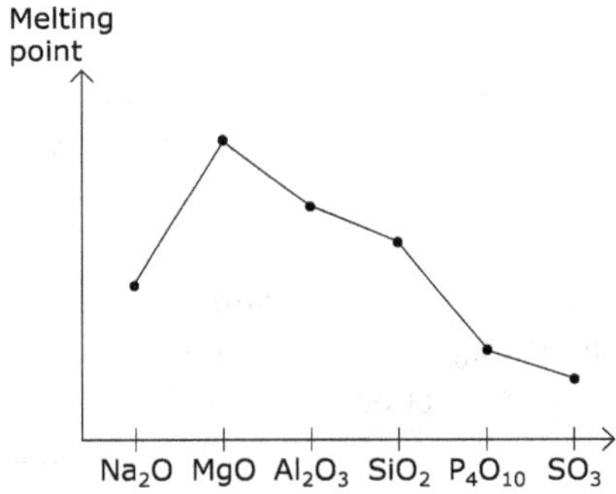

(3) Chemical properties of Period 3 oxides in water
— Ionic metal oxide dissolves (if it does) to give an alkaline solution upon hydrolysis. This is because of the electron-rich O^{2-} ion, which can extract a H^+ ion from a water molecule, forming OH^- ions.
— If the oxide is insoluble in water, it is due to strong ionic bond (MgO and Al_2O_3) or covalent bond (SiO_2). As such, the energy that is released during hydration cannot compensate for the energy that is required to break up the ionic lattice or the covalent bond.

(Continued)

(Continued)

— Non-metal covalent oxides, P_4O_{10} and SO_3, react with H_2O to give a weak acidic solution because the electron-deficient P or S atom is able to attract the lone pair of electrons of a H_2O molecule. The attractive force weakens the O–H bond and causes it to break.

— Na_2O reacts with H_2O to form aqueous NaOH where pH ≈ 12:

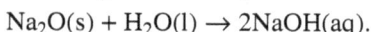

$$Na_2O(s) + H_2O(l) \rightarrow 2NaOH(aq).$$

MgO is sparingly soluble, and gives a solution where pH ≈ 9:

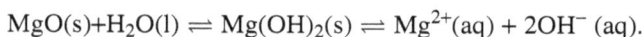

$$MgO(s)+H_2O(l) \rightleftharpoons Mg(OH)_2(s) \rightleftharpoons Mg^{2+}(aq) + 2OH^-(aq).$$

$Al_2O_3$and SiO_2 are both insoluble in H_2O. As a result, the mixture has a pH = 7.

P_4O_{10} and SO_3 react with water to give oxoacids with pH ≈ 2:

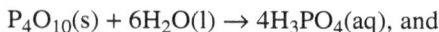

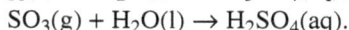

$$P_4O_{10}(s) + 6H_2O(l) \rightarrow 4H_3PO_4(aq), \text{ and}$$
$$SO_3(g) + H_2O(l) \rightarrow H_2SO_4(aq).$$

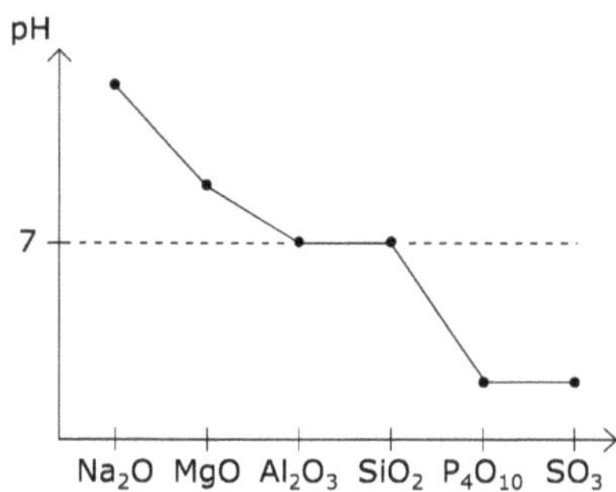

(4) Reaction of the Period 3 oxides with an acid and a base

— Ionic metal oxides (Na_2O and MgO) react with an acid because the electron-rich O^{2-} ion is able to react with the H^+ ion.

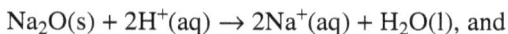

$$Na_2O(s) + 2H^+(aq) \rightarrow 2Na^+(aq) + H_2O(l), \text{ and}$$
$$MgO(s) + 2H^+(aq) \rightarrow Mg^{2+}(aq) + H_2O(l).$$

— Al_2O_3 is amphoteric, i.e., able to react with both an acid and a base, because the high charge density of the Al^{3+} ion is able to attract the OH^- ion, whereas the electron-rich O^{2-} ion is able to react with the H^+ ion.

(Continued)

(Continued)

$$Al_2O_3(s) + 6H^+(aq) \rightarrow 2Al^{3+}(aq) + 3H_2O(l)$$
$$Al_2O_3(s) + 2OH^-(aq) + 3H_2O(l) \rightarrow 2\,[Al(OH)_4]^-(aq).$$

— Non-metal covalent oxides, SiO_2, P_4O_{10}, and SO_3, react with a base because the electron-deficient Si, P, or S atom is able to attract the electron-rich OH^- ion.

$$SiO_2(s) + 2OH^-(aq) \rightarrow SiO_3^{2-}(aq) + H_2O(l);$$
$$P_4O_{10}(s) + 12OH^-(aq) \rightarrow 4PO_4^{3-}(aq) + 6H_2O(l);\ \text{and}$$
$$SO_3(g) + 2OH^-(aq) \rightarrow SO_4^{2-}(aq) + H_2O(l).$$

Do you know?

— In a nutshell, an acidic solution may be due to one of the following:
(1) non-metal chloride;
(2) non-metal oxide;
(3) highly charge density cation; or
(4) the conjugate acid of a weak base.

— A basic solution may be due to one of the following:
(1) a soluble metal oxide or hydroxide, or
(2) the conjugate base of a weak acid.

5. (a) Give the formulae of the chlorides of the elements of Period 3.

Explanation:

$NaCl$; $MgCl_2$; $SiCl_4$; PCl_3 or PCl_5; and SCl_2.

(b) Calculate the percentage by mass of chlorine in the chloride of silicon.

Explanation:

Molar mass of $SiCl_4 = 28.1 + 4(35.5) = 170.1\,g\,mol^{-1}$.
Percentage by mass of Cl in $SiCl_4 = \frac{142}{170.1} \times 100 = 83.5\%$.

(c) (i) Draw a dot-and-cross diagram to show the bonding in the chloride of silicon.

Explanation:

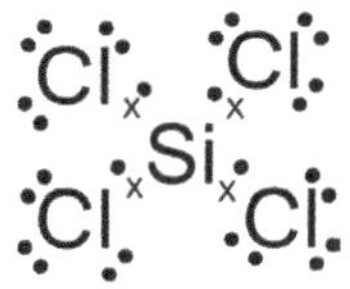

(ii) Draw the shape of this molecule. Explain your answer in terms of the Valence Shell Electron Pair Repulsion (VSEPR) theory.

Explanation:

No. of regions of electron densities = 4.

According to the VSEPR theory, as electron pairs would spread out as far apart as possible to minimize inter-electronic repulsion, the electron pair geometry (EPG) is tetrahedral. Since there is no other lone pair of electrons, the molecular geometry (MG) is also tetrahedral.

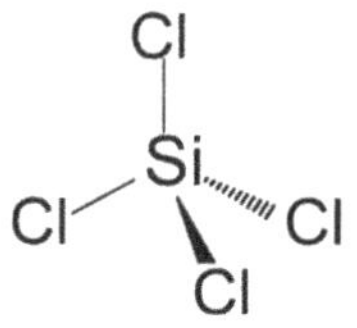

(iii) State the shape of a molecule of $AlCl_3$ and explain why it is different from that of the chloride of silicon.

Explanation:

The molecular geometry of $AlCl_3$ is trigonal planar. This is because there are three regions of electron densities around the Al atom since Al atom has only three valence electrons and hence, it can only form a maximum of three single bonds.

(iv) Give an equation for the reaction of the chloride of silicon with cold water.

Explanation:

$$SiCl_4(l) + H_2O(l) \rightarrow SiCl_3(OH)(s) + HCl(aq).$$

Do you know?

— Cold water does not allow $SiCl_4$ to be completely hydrolyzed to form SiO_2. This is because the hydrolysis of $SiCl_4$ is exothermic, like that of PCl_5 which has been discussed previously. Cold water prevents the futher hydrolysis of the $SiCl_3(OH)(s)$ intermediate.

(v) How does the behavior of carbon tetrachloride with cold water compare with that in part (*iv*)? Explain any differences.

Explanation:

In the first place, CCl_4 is not even soluble in water, so it cannot hydrolyze in water. In addition, the water molecule cannot attack the carbon center because of steric effect posed by the bulky chlorine atoms. The lack of vacant low-lying orbitals to form bonds with the water molecule is another reason why CCl_4 does not hydrolyze in water.

6. (a) Study the table of ionization energies (I.E.) below and answer the questions which follow.

Ionization energy/kJ mol^{-1}	1st	2nd	3rd	4th
Sodium	494	4560	6940	9540
Magnesium	736	1450	7740	10500
Aluminum	577	1820	2740	11600

Explain the *relative* magnitudes of the following:

(i) the 1st I.E. of sodium and magnesium;

Explanation:

Electronic configuration: Na [Ne]$3s^1$ Mg [Ne]$3s^2$

Mg has a higher 1st I. E. than Na because of the greater effective nuclear charge acting on the valence electrons of the Mg atom.

Do you know?

— The inter-electronic repulsion among the pair of electrons in the 3*s* orbital of Mg does not cause Mg to have a lower I.E. than Na. This is because inter-electronic repulsion between a pair of electrons in an *s* orbital is not as significant as that in a *p* orbital or *d* orbital, due to the bigger size of the *s* orbital as compared to the other two orbitals.

(ii) the 1st I. E. of magnesium and aluminium;

Explanation:

Electronic configuration: Mg [Ne]$3s^2$ Al [Ne]$3s^23p^1$

Al has a lower 1st I.E. than Mg because the electron that is removed from Al comes from a higher-energy *p* orbital while that from Mg comes from an *s* orbital. Since the *p* electron is farther away from the nucleus, it is less strongly attracted by the nucleus.

(i) the 2nd I.E. of sodium and magnesium;

Explanation:

Electronic configuration: Na^+ $1s^22s^22p^6$ Mg^+ $[Ne]3s^1$

Na^+ has a much higher 2nd I.E. than Mg^+ because the electron that is being removed from Na^+ comes from a lower-energy principal quantum shell that is closer to the nucleus than the electron that is removed from the 3*s* orbital of Mg^+. Thus, the electron that is to be removed from Na^+ is more strongly attracted.

(iv) the 2nd I.E. of magnesium and aluminium;

Explanation:

Electronic configuration: Mg^+ $[Ne]3s^1$ Al^+ $[Ne]3s^2$

Al^+ has a higher 1st I.E. than Mg^+ because of the greater effective nuclear charge acting on the valence electrons of the Al^+ ion.

(v) the 3rd and 4th I.E.of aluminum.

Explanation:

Electronic configuration: Al^{2+} $[Ne]3s^1$ Al^{3+} $1s^22s^22p^6$

The 4th I.E. of Al is much higher than the 3rd I.E. because the next electron that is to be removed from Al^{3+} comes from a lower-energy principal quantum shell that is closer to the nucleus than the electron that is removed from the 3*s* orbital of Al^{2+}. Thus, the next electron that is to be removed from Al^{3+} is more strongly attracted. In another perspective, since $A1^{2+}$ has more electrons than $A1^{3+}$, the inter-electronic repulsion of $A1^{2+}$ is greater than $A1^{3+}$. As both species have the same nuclear charge, the valence electrons of $A1^{3+}$ is more strongly held by the nucleus, accounting for a higher 4th I.E. than the 3rd I.E.

(b) Consider the electron affinities for oxygen given below.

Electron affinity/kJ mol^{-1}	1st	2nd
	−142	+844

(i) Write equations representing the changes to which the 1st and 2nd electron affinities of oxygen relate.

Explanation:

1st electron affinity of O atom: $O(g) + e^- \rightarrow O^-(g)$, and
2nd electron affinity of O atom: $O^-(g) + e^- \rightarrow O^{2-}(g)$.

(ii) Explain the relative magnitudes of the 1st and 2nd electron affinities of oxygen.

Explanation:

The 2nd electron affinity of the O atom is endothermic while the 1st electron affinity is exothermic, because when the 2nd electron approaches the negatively charged ion, the inter-electronic repulsion between the incoming electron and the extra electron is much greater than the attractive force the nucleus acts on this incoming electron. Hence, work must be done to "force" this 2nd electron into the ion.

Do you know?

— Electron affinity measures the attractive force between the incoming electron and the nucleus. The stronger the attractive force, the more exothermic would the electron affinity be.

— Since electron affinity is a measure of the attractive force the nucleus has for the incoming electron, then the electron affinity should be less

(*Continued*)

(*Continued*)

exothermic down the group. This is because as we go down a group, no doubt the nuclear charge increases but the shielding effect also increases as the number of inner core electrons increases. Hence, the net attractive force on the valence electrons, including the incoming electron or outgoing electron decreases.

— The factors that affect electron affinity are very similar to those for factors that affect I.E., which include ENC (e.g., O atom versus N atom), inter-electronic repulsion, charge of the species (e.g., O atom versus O^-), and distance from the nucleus (down the group).

Why is the 1st electron affinity for the S atom ($-200.4\,kJ\,mol^{-1}$) more exothermic than that of the O atom ($-142\,kJ\,mol^{-1}$)?

A: The incoming electron that is being brought into the S atom would be less strongly attracted by the nucleus as the valence shell is farther away from the nucleus. But the valence shell of an O atom is much smaller than that of the S atom. Hence, the inter-electronic repulsion between the incoming electron and those electrons that are already in the valence shell is much greater for the O atom than for the S atom. Thus, this factor actually overshadows the distance from the nucleus factor and causes the electron affinity for the O atom to be less exothermic than that for the S atom. Similarly, if you look at the 1st electron affinity of F ($-328.2\,kJ\,mol^{-1}$), it is also less exothermic than that of Cl ($-349.0\,kJ\,mol^{-1}$).

(iii) Given the endothermic nature of the 2nd electron affinity of oxygen, comment briefly on the thermodynamic stability of ionic metal oxides.

Explanation:

Since more energy is needed to form the O^{2-}(g) ion, the metal oxide formed should be thermodynamically less stable. But a doubly charged anion also results in a more exothermic lattice energy, thus lowering the energy level of the metal oxide, hence increase the thermodynamic stability of the compound.

CHAPTER 10

CHEMISTRY OF GROUPS 2 AND 17

What do you know regarding the Group 17 elements and their chemistry?

(1) Elements of concern: F, Cl, Br, I

All exist as simple discrete diatomic molecules.

Elements from Cl to I can exist in variable oxidation states and the maximum oxidation number corresponds to the number of valence electrons the atom has, i.e., +7. This is because Cl to I atoms possess vacant low-lying *d* orbitals which allow the expansion of the octet configuration.

(2) Trends in physical properties of the elements

(i) *Volatility, melting point, and boiling point*

— Since the elemental form consists of simple discrete non-polar molecules, the number of electrons increases from F_2 to I_2. As a result, the electron cloud becomes more polarizable, and the strength of instantaneous dipole–induced dipole (id–id) interaction increases. This results in lower volatility, lower saturated vapor pressure, more endothermic enthalpy change of vaporization, or higher melting and boiling points.

(ii) *Solubility in water*

— Fluorine oxidizes water vigorously to form oxygen:

$$2F_2(g) + 2H_2O(l) \rightarrow 4HF(aq) + O_2(g).$$

— Chlorine disproportionates partially in water to form hydrochloric and chloric (I) acids:

$$Cl_2(g) + H_2O(l) \rightleftharpoons HCl(aq) + HClO(aq).$$

HClO is responsible for the bleaching and disinfecting action of chlorine water.

— Both Br_2 and I_2 are sparingly soluble in water. This is because the id–id interaction between the non-polar halogen molecule and water molecule does not release a sufficient amount of energy to compensate for the energy required to overcome the strong hydrogen bonding between water molecules.

(iii) *Solubility in organic solvent*

— Group 17 molecules are non-polar, hence they are highly soluble in non-polar organic solvents. This is because the id–id interaction between the solute and solvent molecules release a sufficient amount of energy to compensate for the energy required to overcome the intermolecular forces within the solute and solvent molecules.

(iv) *Bond energy*

— Bond energy decreases down the group. This is because the orbital involved in the covalent bond formation is bigger and hence more diffused. The overlap is less effective, resulting in weaker covalent bond being formed.

(v) *Electron affinity*

— Electron affinity becomes less exothermic down the group. This is because the extra electron experiences weaker attractive force by the nucleus due to an increase in the distance of the valence shell from the nucleus.

(3) Trends in chemical properties of the elements

Down the group, the ability to gain electrons decreases. This is because the extra electron sits in a valence shell that is farther away from the nucleus; hence, it is less strongly attracted. As a result, the oxidizing power (the ability to be reduced) decreases down the group, which is reflected in less positive E^θ values.

$$F_2 + 2e^- \rightleftharpoons 2F^-, \quad E^\theta = +2.8\ V;$$
$$Cl_2 + 2e^- \rightleftharpoons 2Cl^-, \quad E^\theta = +1.36\ V;$$
$$Br_2 + 2e^- \rightleftharpoons 2Br^-, \quad E^\theta = +1.09\ V;\ \text{and}$$
$$I_2 + 2e^- \rightleftharpoons 2I^-, \quad E^\theta = +0.54\ V.$$

The decrease in oxidizing power of the Group 17 elements is mediated in various reactions as follows:

(i) *Displacement reaction*

As the oxidizing power decreases down the group, halogens would be able to displace the less reactive halide below it in the group. Example:

$$Cl_2(g) + 2Br^-(aq) \rightarrow 2Cl^-(aq) + Br_2(aq).$$
$$E^\theta_{cell} = E^\theta_{Red} - E^\theta_{Ox} = +1.36 - (+1.09) = +0.27\ V.$$

The addition of a non-polar organic solvent such as hexane would result in a reddish-brown layer containing Br_2 in an organic solvent being observed.

(ii) *Reaction with* Fe^{2+}

I_2 does not react with Fe^{2+} but the rest of the halogens above iodine do:

$$I_2(s) + 2Fe^{2+}(aq) \rightarrow 2I^-(aq) + 2Fe^{3+}(aq).$$

$$E^\theta_{cell} = +0.54 - (+0.77) = -0.23 \text{ V}.$$

The non-reaction between I_2 and Fe^{2+} can be verified through the calculation of E^θ_{cell}.

(iii) *Reaction with thiosulfate*

As the oxidizing power decreases down the group, halogens would oxidize the sulfur atom of the thiosulfate to a different oxidation state. The greater the oxidizing power, the higher the oxidation number of the sulfur in the final product:

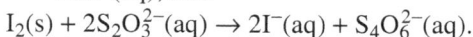

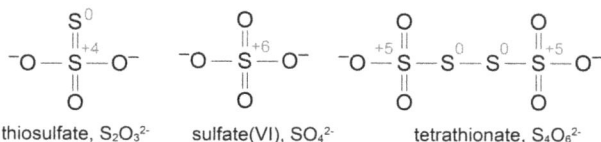

$$4Cl_2(g) + S_2O_3^{2-}(aq) + 5H_2O(l) \rightarrow 8Cl^-(aq) + 2SO_4^{2-}(aq) + 10H^+(aq);$$

$$4Br_2(l) + S_2O_3^{2-}(aq) + 5H_2O(l) \rightarrow 8Br^-(aq) + 2SO_4^{2-}(aq) + 10H^+(aq); \text{ and}$$

$$I_2(s) + 2S_2O_3^{2-}(aq) \rightarrow 2I^-(aq) + S_4O_6^{2-}(aq).$$

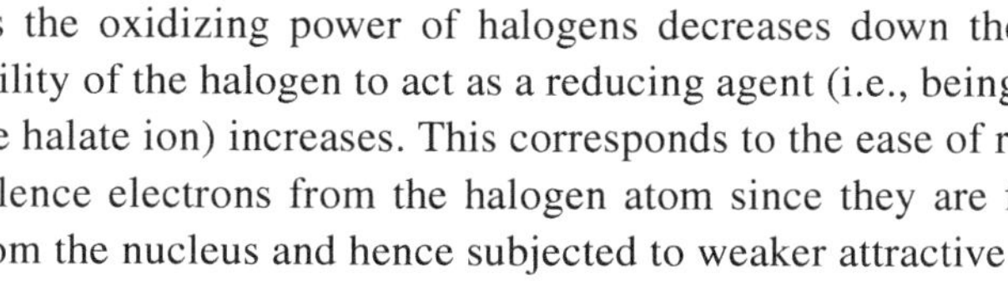

thiosulfate, $S_2O_3^{2-}$ sulfate(VI), SO_4^{2-} tetrathionate, $S_4O_6^{2-}$

(iv) *Reaction with alkaline*

As the oxidizing power of halogens decreases down the group, the ability of the halogen to act as a reducing agent (i.e., being oxidized to the halate ion) increases. This corresponds to the ease of removing the valence electrons from the halogen atom since they are farther away from the nucleus and hence subjected to weaker attractive force by the nucleus.

Halogen disproportionates to form the halide and halate(I) ions at low temperatures:

$$X_2 + 2OH^-(aq) \rightarrow \underbrace{X^-(aq)}_{\text{halide ion}} + \underbrace{XO^-(aq)}_{\text{halate(I) ion}} + H_2O(l).$$

Halate(I) ion can further disproportionate to the halide and halate(V) ions at higher temperatures:

$3XO^-(aq) \rightarrow$	$2X^-(aq) +$	$XO_3^-(aq)$.
halate(I) ion	halide ion	halate(V) ion

When alkali reacts with:	Major halate product obtained at:		
	0°C	15°C	70°C
Cl_2	ClO^-	ClO^-	ClO_3^-
Br_2	BrO^-	BrO_3^-	BrO_3^-
I_2	← mostly IO_3^- →		

Thus, as the temperature increases for an alkaline solution of Cl_2, the smell of the Cl_2 disappears not because the Cl_2 has evaporated. It is in fact due to the further disproportionation of the ClO^- to ClO_3^-.

(v) *Reaction with hydrogen gas*

As the oxidizing power of halogens decreases down the group, the rate of the reaction between the halogen and hydrogen gas decreases down the group as well:

Element	Reaction with hydrogen
F_2	Explosively in the dark
Cl_2	Explosively in the presence of light
Br_2	React at 300°C in the presence of Pt catalyst
I_2	Strong heating and catalyst needed; reaction is incomplete

The X–X bond becomes weaker down the group and is thus easier to break , which cannot be used to account for the above observation. And the main reason to account for the above observation is because the H–X bond becomes weaker as one progresses down the group. Thus, the formation of HX becomes less favored.

(4) Trends of the properties of hydrogen halides

(i) *Volatility*

The boiling point of HX varies according to the order: HF > HI > HBr > HCl. The highest boiling point of HF is ascribed to hydrogen bonding, although the bond polarity decreases from HCl to HBr to HI, meaning that the permanent dipole–permanent dipole interaction would also decrease in the same direction. The increase in boiling point from HCl to HBr to HI is actually due to an increase in the number of electrons, hence enhancing the strength of id–id interaction.

(ii) *Bond energy*
The H–X bond becomes weaker down the group as the overlap of the atomic orbitals becomes less effective. As a result, the bond energy becomes less endothermic. Hence, HX becomes more thermally unstable down the group.

(iii) *Acidity of HX*
Due to the weaker H–X covalent bond, the strength of the acid increases down the group. In fact, HCl, HBr, and HI all fully dissociate in water.

(iv) *Reducing power of HX*
HX becomes a stronger reducing agent, i.e., likely to undergo oxidation, down the group. This is related to the ease of removing electrons from the valence shell of the X atom. This phenomenon is observed in the preparation of HX when concentrated H_2SO_4 acts on sodium halide, NaX:

$$NaCl(s) + H_2SO_4(l) \rightarrow NaHSO_4(s) + HCl(g);$$
$$NaBr(s) + H_2SO_4(l) \rightarrow NaHSO_4(s) + HBr(g)$$
$$2HBr(g) + H_2SO_4(l) \rightarrow SO_2(g) + Br_2(l) + 2H_2O(l); \quad \text{and}$$
$$NaI(s) + H_2SO_4(l) \rightarrow NaHSO_4(s) + HI(g)$$
$$2HI(g) + H_2SO_4(l) \rightarrow SO_2(g) + I_2(s) + 2H_2O(l)$$
$$6HI(g) + H_2SO_4(l) \rightarrow S(s) + 3I_2(s) + 4H_2O(l)$$
$$8HI(g) + H_2SO_4(l) \rightarrow H_2S(g) + 4I_2(s) + 4H_2O(l).$$

The stronger the reducing power of HI, the lower the oxidation state the S atom of H_2SO_4 would be reduced to, i.e., to zero or –2.

(1) Distinguishing test for halides

	Type of halide	$Cl^-(aq)$	$Br^-(aq)$	$I^-(aq)$
Step 1	Add $AgNO_3(aq)$ to the test tube containing the halide	White ppt of AgCl formed	Cream ppt of AgBr formed	Yellow ppt of AgI formed
Step 2a	To the ppt in Step 1, add dilute $NH_3(aq)$	Ppt soluble in dilute $NH_3(aq)$	Ppt insoluble in dilute $NH_3(aq)$	Ppt insoluble indilute $NH_3(aq)$
Step 2b	To the ppt in Step 1, add conc. NH_3	Ppt soluble in conc. NH_3	Ppt soluble in conc. NH_3	Ppt insoluble in conc. NH_3

— The solubility of the AgX decreases from left to right: AgCl > AgBr > AgI.
— As AgCl is more soluble, a suspension of insoluble AgCl would contain the highest $[Ag^+]$. As a result, when aq. NH_3 is added, there would be sufficient $[Ag(NH_3)_2]^+$ formed such that the position of the $AgCl(s) \rightleftharpoons Ag^+(aq) + Cl^-(aq)$ equilibrium would be shifted to the right, causing AgCl to dissolve.
— For AgBr, the $[Ag^+]$ in the suspension is too low such that concentrated aq. NH_3 is required to solubilize the AgBr.

(2) Some uses of Group 17 elements and their compounds
- Fluorine is used in the making of chlorofluorocarbon (CFC) compounds and PTFE (poly(tetrafluoroethene)).
- Chlorine is used as a disinfectant in swimming pools and the making of PVC (polyvinyl chloride).
- Bromine is used to make AgBr used in photographic films.

1. (a) Outline one industrial process for the manufacture of chlorine, stating the starting materials.

Explanation:

Chlorine is produced through electrolysis of brine solution, which is actually concentrated NaCl(aq), using a diaphragm cell.

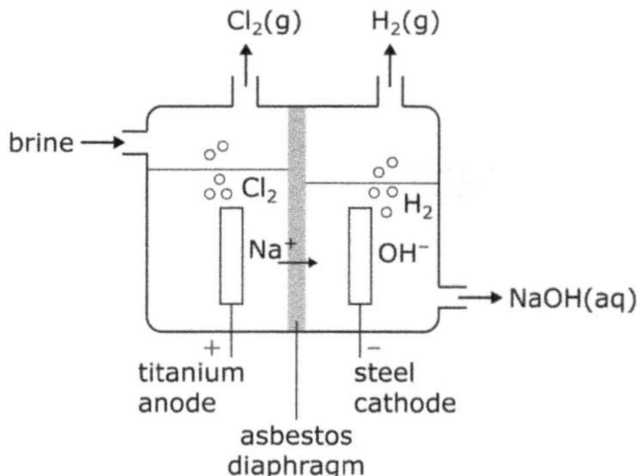

The brine solution is introduced into the anode compartment and passed through a semi-permeable asbestos diaphragm into the cathode compartment. At the titanium anode, the Cl^- ion is oxidized to Cl_2: $2Cl^-(aq) \rightarrow Cl_2(g) + 2e^-$.

While at the steel electrode, the following reaction occurs:

$$2H_2O(l) + 2e^- \rightarrow H_2(g) + 2OH^-(aq).$$

As a result, the cathode compartment becomes saturated with NaOH(aq) which is important for the production of bleach, when mixed with the Cl_2 that is produced at the anode:

$$2NaOH(aq) + Cl_2(aq) \rightarrow NaClO(aq) + NaCl(aq) + H_2O(l).$$

Both the Cl_2 and H_2 gases are not allowed to mix as the formation of HCl(g) is very explosive in nature. In addition, the level of solution in the anode compartment is always kept at a higher level to prevent back-mixing of the solution from the cathode compartment.

Do you know?

— Why does the Cl^- undergo oxidation in the anode compartment, instead of the H_2O or OH^- ion? (Refer to Chapter 8)

— Why does the H_2O undergo reduction in the cathode compartment, instead of the Na^+ or H^+ ion? (Refer to Chapter 8)

(b) Name any one commercially important compound which is made directly or indirectly from chlorine and state its use.

Explanation:

Chlorine is important for the making of chloroethene ($CHCl{=}CH_2$), which is critical for the making of poly(chloroethene) or PVC.

(c) Chlorine reacts slightly with water. The reversible reaction can be represented by the equation: $Cl_2 + H_2O \rightleftharpoons HCl + HClO$.

(i) What is the oxidation state of chlorine in HClO?

Explanation:

The oxidation state of chlorine in HClO is +1.

Do you know?

— HClO is known as chloric(I) acid as the chlorine atom has a +1 oxidation state. The ClO^- is known as chlorate(I) ion.

Q Why did the reaction $Cl_2 + H_2O \rightleftharpoons HCl + HClO$ not go to completion but instead, an equilibrium is established?

A: If you calculate the E^θ_{cell} for the reaction, $Cl_2 + H_2O \rightleftharpoons HCl + HClO$:

$$2HClO + 2H^+ + 2e^- \rightleftharpoons Cl_2 + 2H_2O, \quad E^\theta = +1.64\text{ V}, \quad \text{and}$$
$$Cl_2 + 2e^- \rightleftharpoons 2Cl^-, \quad E^\theta = +1.36\text{ V}.$$

$E^\theta_{cell} = E^\theta_{Red} - E^\theta_{Ox} = 1.36 - (+\ 1.64) = -0.28$ V.

Since the $E^\theta_{cell} < 0$, the reaction is actually thermodynamically non-spontaneous under standard conditions.

Q So, if the reaction is thermodynamically non-spontaneous under standard conditions, why did it still happen?

A: Well, it happened because the conditions that made it happen was not under standard conditions. That is why the reaction did not go to completion.

(ii) Give the name of the reaction.

Explanation:

Since the chlorine undergoes both oxidation and reduction, it is a disproportionation reaction.

(i) Write two ionic half-equations for this process.

Explanation:

Oxidation half-equation: $Cl_2 + 2H_2O \rightarrow 2ClO^- + 4H^+ + 2e^-$.
Reduction half-equation: $Cl_2 + 2e^- \rightarrow 2Cl^-$.

(d) Chlorine is widely used to disinfect water. The non-ionized acid HClO, formed in the above reaction, is a much better disinfectant than the ClO^- ion. HClO is a weak acid:

$$HClO(aq) \rightleftharpoons H^+(aq) + ClO^-(aq).$$

(i) In which direction should the pH of the solution be adjusted so as to increase the disinfecting power? Explain your answer.

Explanation:

Since HClO is a better disinfectant than ClO^-, the pH of the solution should be decreased to shift the position of equilibrium of the reaction, $HClO(aq) \rightleftharpoons H^+(aq) + ClO^-(aq)$, to the left. This is because at a lower pH, $[H^+(aq)]$ increases; according to Le Chatelier's Principle, the reaction must respond in such a way to decrease the $[H^+(aq)]$. Hence, the position of equilibrium would shift to the left.

Q Why is HClO a better disinfectant than ClO^-?

A: Both HClO and ClO^- work as disinfectants by oxidation. To understand why HClO is a better disinfectant than ClO^-, we can look at their standard reduction potential:

$$2HClO + 2H^+ + 2e^- \rightleftharpoons Cl_2 + 2H_2O, \quad E^\theta = +1.64\text{ V}, \text{ and}$$
$$ClO^- + H_2O + 2e^- \rightleftharpoons Cl^- + 2OH^-, \quad E^\theta = +0.90\text{ V}.$$

From the E^θ values, it can be seen that HClO undergoes reduction more readily than ClO^-, hence it is a stronger oxidizing agent.

(ii) Give one adverse effect which might follow if the pH were adjusted too far in this direction.

Explanation:

If the pH is too low, then [HClO(aq)] would increase too much. Then, according to Le Chatelier's Principle, the position of equilibrium of the reaction, $Cl_2 + H_2O \rightleftharpoons HCl + HClO$, would shift to the left. Hence, Cl_2 gas would be produced.

Do you know?

— To prevent chlorine gas from escaping, it is preferably dissolved in NaOH as the alkaline medium shifts the position of equilibrium of $Cl_2 + H_2O \rightleftharpoons HCl + HClO$, toward the side of the formation of HClO, i.e., $Cl_2 + 2OH^- \rightleftharpoons Cl^- + ClO^- + H_2O$.

2. Bromine occurs as bromide ions in low concentration in seawater. It is extracted commercially by the following series of steps:

   A Chlorine is passed into acidified seawater and bromine is then removed in a stream of air. The concentration of bromine in the gas phase is too low at this stage for it to be condensed out efficiently.

   B Therefore, it is converted to hydrogen bromide by reaction with added sulfur dioxide and a small excess of water vapor:

   $$Br_2(g) + SO_2(g) + 2H_2O(g) \rightarrow 2HBr(g) + H_2SO_4(g).$$

   A moderately concentrated, aqueous mixture of hydrobromic and sulfuric acids is obtained when the vapor is cooled and condensed.

   C The concentrated mixture of acids is treated with chlorine and steam, producing a vapor from which liquid bromine is condensed on cooling. The sulfuric acid remains in the solution.

   D The crude bromine is purified by fractional distillation.

   (a) Explain, in terms of the relevant standard electrode potentials, the reactions taking place in steps A and B. (Assume $SO_2 + H_2O$ is H_2SO_3.)

Explanation:

Step **A**:

$$Cl_2 + 2e^- \rightleftharpoons 2Cl^-, \quad E^\theta = +1.36\text{ V, and}$$
$$Br_2 + 2e^- \rightleftharpoons 2Br^-, \quad E^\theta = +1.09\text{ V.}$$

The redox reaction: $Cl_2 + 2Br^- \rightarrow 2Cl^- + Br_2$.
Calculating $E^\theta{}_{cell} = E^\theta{}_{Red} - E^\theta{}_{Ox} = 1.36 - (+1.09) = +0.27$ V.
Since $E^\theta{}_{cell} > 0$, the redox reaction is thermodynamically spontaneous under standard conditions.

Step B:

$$SO_4^{2-} + 4H^+ + 2e^- \rightleftharpoons SO_2 + 2H_2O, \quad E^\theta = +0.17\text{ V, and}$$
$$Br_2 + 2e^- \rightleftharpoons 2Br^-, \quad E^\theta = +1.09\text{ V.}$$

The redox reaction: $Br_2(g) + SO_2(g) + 2H_2O(g) \rightarrow 2HBr(g) + H_2SO_4(g)$.
Calculating $E^\theta{}_{cell} = E^\theta{}_{Red} - E^\theta{}_{Ox} = 1.09 - (+0.17) = +0.92$ V.
Since $E^\theta{}_{cell} > 0$, the redox reaction is thermodynamically spontaneous under standard conditions.

Do you know?

— If you calculate the $E^\theta{}_{cell}$ value for $Br_2 + 2Cl^- \rightarrow 2Br^- + Cl_2$, you would find that it is a negative value. The reaction is thermodynamically non-spontaneous. Hence, Br_2 cannot oxidize Cl^-! This is in conjunction with the fact that the oxidizing power of the Group 17 elements decreases down the group.

$$E^\theta{}_{cell} = E^\theta{}_{Red} - E^\theta{}_{Ox} = 1.09 - (+1.36) = -0.27\text{ V.}$$

(b) Why does liquid bromine vaporize easily?

Explanation:

Bromine consists of non-polar diatomic molecules with weak instantaneous dipole–induced dipole interaction. Hence, it has a low boiling point.

Do you know?

— The physical state of the substance is an indication of the strength of the bonds that hold the particles together. Cl_2 is a gas, Br_2 is a liquid, while I_2 is a solid, all at room temperature. These show that the instantaneous dipole–induced dipole interaction increases down the group due to the increase in the number of electrons the non-polar diatomic molecule has.

(c) Explain why, in Step **B**, a large volume of gaseous hydrogen bromide dissolves in a small volume of water, and give the chemical species present in aqueous hydrobromic acid.

Explanation:

HBr is a strong acid that fully dissociates in water. This accounts for its high solubility in water. The chemical species that are present in HBr(aq) are H_3O^+(aq) and Br^-(aq).

Do you know?

— The acid strength increases from HCl to HBr to HI. This is due to the weakening of the H–X bond down the group. As the orbital that the halogen, X, used to form the bond with the H atom is bigger, it is more diffused and hence the orbital overlap of the H–X bond is less effective down the group.

(d) Describe a test for the detection of bromide ions in aqueous solution.

Explanation:

The addition of $AgNO_3(aq)$ into a solution containing $Br^-(aq)$ would result in the formation of a cream ppt of AgBr, which is insoluble in $NH_3(aq)$.

Do you know?

— Alternatively, you can add an oxidizing agent that is able to convert Br^- into orange Br_2. Then, add some hexane and shake. If the hexane turns orange-brown, it is a confirmation that you have Br_2.

What do you know regarding the Group 2 elements and their chemistry?

(1) Elements of concern: Be, Mg, Ca, Sr, Ba
The maximum oxidation state of the element corresponds to the number of valence electrons the atom has, i.e., +2.

(2) Trends in chemical reactivity of the elements
Down the group, the ability to lose electrons increases. This is because the valence electrons are farther away from the nucleus, hence they are less strongly attracted. As a result, the reducing power (the ability to be oxidized) increases down the group, as reflected in the more negative E^θvalues.

$$Mg^{2+} + 2e^- \rightleftharpoons Mg, \quad E^\theta = -2.38 \text{ V};$$
$$Ca^{2+} + 2e^- \rightleftharpoons Ca, \quad E^\theta = -2.87 \text{ V};$$
$$Sr^{2+} + 2e^- \rightleftharpoons Sr, \quad E^\theta = -2.89 \text{ V}; \text{ and}$$
$$Ba^{2+} + 2e^- \rightleftharpoons Ba, \quad E^\theta = -2.90 \text{ V}.$$

The increase in reducing power of the Group 2 elements is mediated in various reactions as follows:

(i) *Reaction with water*

Element	Reaction with water	Equation
Mg	Very slow reaction with cold water	$M(s) + 2H_2O(l) \rightarrow$
Ca	Readily reacts with cold water	$M(OH)_2(aq) + H_2(g)$
Sr	Vigorous reaction with cold water	
Ba	Very vigorous reaction with cold water	Note: $Mg(OH)_2$ is insoluble.

(ii) *Reaction with oxygen*

Element	Reaction with oxygen	Equation
Mg	Reacts above its melting point	$2M(s) + O_2(g) \rightarrow$
Ca	Reacts upon heating	$2MO(s)$
Sr	Reacts upon heating	
Ba	Reacts spontaneously without heating	

(3) Chemical properties of Group 2 oxides

— BeO is an amphoteric oxide, i.e., it is able to react with both an acid and a base. This is because the high charge density of the Be^{2+} ion is able to attract the OH^- ion, whereas the electron-rich O^{2-} ion is able to react with the H^+ ion:

$$BeO + 2H^+ \rightarrow Be^{2+} + H_2O$$
$$BeO + 2OH^- + H_2O\ (1) \rightarrow [Be(OH)_4]^{2-}.$$

— The other Group 2 oxides are ONLY basic in nature due the electron-rich O^{2-} ion.

(4) Thermal stability of Group 2 compounds

The thermal stability of Group 2 compounds increases down the group, i.e., a higher temperature is required for decomposition to take place. This is because as we go down the group, the charge of the cation is the same but the cationic radius increases down the group. As a result, the charge density decreases down the group. When the charge density decreases, the polarizing power also decreases, thus the electron cloud of the anion is distorted to a lesser extent. Hence, there is less weakening of the intramolecular covalent bonds within the anion, which thus requires a higher temperature to break. In addition, thermal decomposition is also driven by an increase in the entropy of the system at a higher temperature due to the formation of gas particles such as CO_2 or $NO_2 + O_2$ or H_2O.

(i) *Decomposition of nitrate*

$$M(NO_3)_2(s) \rightarrow MO(s) + 2NO_2(g) + \tfrac{1}{2}\,O_2(g).$$

(ii) *Decomposition of carbonate*

$$MCO_3(s) \rightarrow MO(s) + CO_2(g).$$

(iii) *Decomposition of hydroxide*

$$M(OH)_2(s) \rightarrow MO(s) + H_2O(g).$$

(5) Some uses of Group 2 elements and their compounds

- Magnesium oxide: because of its high melting point and low reactivity, it is mainly used as a refractory material in furnace linings.
- Magnesium hydroxide: used in "milk of magnesia" for the alleviation of constipation and treatment of acid indigestion.
- Calcium oxide (known as lime or quicklime): when mixed with water, calcium hydroxide (slaked lime) is produced, which can be used for reducing soil acidity in agriculture and in the production of mortar and plaster.
- Calcium carbonate: has primary uses in the construction industry as building materials.
- Barium sulfate: used as a radiocontrast agent for the X-ray imaging of the gastrointestinal tract.

3. (a) Explain why alkaline solutions result when the oxides of the *s* block elements dissolve in water.

Explanation:

When the oxides of the *s* block elements dissolve in water, it releases the O^{2-} ion which reacts with the H_2O molecule to give OH^-:

$$O^{2-} + H_2O \rightarrow 2OH^-.$$

Q Why isn't MgO soluble in water to give an alkaline solution?

A: This is because the hydration energies released by the Mg^{2+} and O^{2-} are insufficient to compensate for the energy that is needed to break up the ionic lattice.

(b) Most of these oxides dissolve readily in water. However, the oxides of beryllium and magnesium have very low solubility in water. Explain in terms of enthalpy changes why this is so.

Explanation:

The oxide of beryllium and magnesium have very low solubility in water because the hydration energies released by the Mg^{2+} or Be^{2+} and O^{2-} are insufficient to compensate for the energy that is needed to break up the ionic lattice.

(c) An aqueous solution of beryllium sulfate is acidic whereas an aqueous solution of magnesium sulfate is almost neutral. Why is there this difference in the behavior of $Be^{2+}(aq)$ and $Mg^{2+}(aq)$?

Explanation:

An aqueous solution of beryllium sulfate is acidic because the $[Be(H_2O)_6]^{2+}$ undergoes appreciable hydrolysis. This is resulted because the charge density of the Be^{2+} is high; the high polarizing power of the Be^{2+} distorts the electron cloud of the H_2O. This weakens the O–H bond of the H_2O molecule to an extent that the $[Be(H_2O)_6]^{2+}$ complex loses a H^+ ion.

$$[Be(H_2O)_6]^{2+} + H_2O \rightleftharpoons [Be(H_2O)_5(OH)]^+ + H_3O^+.$$

Do you know?

— Due to the appreciable acidic hydrolysis of the $[Be(H_2O)_6]^{2+}$ complex, when a carbonate or hydrogencarbonate is added to the beryllium sulfate solution, CO_2 gas evolves. Hence, $BeCO_3$ cannot be synthesized!

— Actually, the charge density of Mg^{2+} is also high enough to cause some acidic hydrolysis to take place. The pH of a solution containing Mg^{2+} ion has a pH $\approx$ 6.5, which is not strong enough to react with carbonate.

Both $BeSO_4$ and $MgSO_4$ are soluble in water, yet the solubility of the Group 2 sulfate(VI) compounds decreases down the group. Is there a reason for it?

A: As we go down the group, the energy required to break up the ionic lattice of the Group 2 sulfate(VI) decreases because of an increase in cationic radius. Hence, the ionic bond strength decreases. But the hydration enthalpy from the Group 2 cation also becomes less exothermic down the group, due to a decrease in charge density. In addition, as SO_4^{2-} is a large anion, the hydration enthalpy provided by this anion is not very exothermic. Overall, as we go down the group, the hydration energies become much less exothermic than the rate of change of the lattice energy. As a result, the hydration enthalpies become even less able to compensate for the energy required to break up the ionic lattice. Therefore, the solubility of the Group 2 sulfates decreases down the group.

(d) Given a sample of solid magnesium chloride contaminated with magnesium carbonate, describe tests you would perform in order to confirm the presence of:

(i) magnesium ions;

Explanation:

To test for Mg^{2+} ions, we can add NaOH(aq) in excess. A white ppt that is insoluble would form in excess NaOH(aq). The white ppt is probably $Mg(OH)_2$.

Do you know?

— All Zn^{2+}, Pb^{2+}, and Al^{3+} ions also form a white ppt with NaOH(aq), but these white ppt are soluble in excess NaOH(aq).

(ii) chloride ions; and

Explanation:

To test for Cl^- ions, we can add $AgNO_3$(aq). A white ppt of AgCl will form.

Do you know?

— When NH_3(aq) is added to AgCl(s), the solid dissolves in NH_3(aq) because the formation of the diamminesilver(I) complex, $[Ag(NH_3)_2]^+$, causes the $[Ag^+]$ to decrease. According to Le Chatelier's Principle, the position of the following equilibrium shifts to the right:

$$AgCl(s) \rightleftharpoons Ag^+(aq) + Cl^-(aq)$$

$$Ag^+(aq) + 2NH_3(aq) \rightarrow [Ag(NH_3)_2]^+.$$

(iii) carbonate ions.

Explanation:

To test for CO_3^{2-} ions, we can add dilute HCl(aq). Effervescene of CO_2 would give a white ppt with $Ca(OH)_2$(aq). The white ppt is $CaCO_3$(s).

Do you know?

— $CO_2(g) + 2OH^-(aq) \rightarrow CO_3^{2-}(aq) + H_2O(l)$.

4. (a) (i) Write an equation for the reaction of barium with water.

Explanation:

$$Ba(s) + 2H_2O(l) \rightarrow Ba(OH)_2(aq) + H_2(g).$$

Do you know?

— The solubility of the Group 2 sulfates(VI) decreases down the group BUT the solubility of the Group 2 hydroxides actually increases down the group. For example, both $Be(OH)_2$ and $Mg(OH)_2$ are insoluble but solubility increases from $Ca(OH)_2$ onward.

— The reason to account for it is: as we go down the group, the energy required to break up the ionic lattice of the Group 2 hydroxides decreases because of an increase in cationic radius. Hence, the ionic bond strength decreases. But the hydration enthalpy from the Group 2 cation also becomes less exothermic down the group, due to a decrease in charge density. In addition, as OH^- is a very small anion, the hydration enthalpy provided by this anion is very exothermic. Overall, as we go down the group, the hydration energies become much more exothermic than the rate of change of the lattice energy. As a result, the hydration enthalpies are more able to compensate for the energy required to break up the ionic lattice. Therefore, solubility increases.

(ii) Would the reaction in *(a)(i)* occur more vigorously or less vigorously than the reaction of calcium with water? Identify one contributing factor and use it to justify your answer.

Explanation:

The reaction in *(a)(i)* occurs more vigorously than the reaction of calcium with water. This is because a Ba atom loses electrons, i.e., undergoes oxidation, more readily than a Ca atom. This arises because the two valence electrons of the Ba atom are farther away from the nucleus, hence they are less strongly attracted as compared to those of Ca.

Do you know?

— The increase in the rate of the reaction of the Group 2 elements with H_2O or O_2 or even with other substances, is a demonstration of the increase in the ability to undergo oxidation down the group. Hence, it is also a reflection of the increase in the reducing power. This is because as we go down Group 2, the valence electrons are farther away from the nucleus; hence, they are less strongly attracted.

— Thus, if the valence electrons are less strongly attracted and are easily lost, this would also mean that the extra electrons that are placed in the valence shell would also be less strongly attracted. So, as the reducing power increases down a group, the oxidizing power MUST decrease. This is what we see both in Groups 2 and 17, i.e., the reducing power increases for Group 2 but the oxidizing power decreases for Group 17. Thus, the main common contributing factor for the seemingly opposite trend is simply because the distance of the valence shell increases down the group. Hence, the valence electrons are less strongly attracted (i.e. easier to be removed during oxidation) or extra electrons put into the valence shell are less strongly attracted during reduction.

(iii) Write an equation for the action of heat on solid barium carbonate.

Explanation:

$$BaCO_3(s) \rightarrow BaO(s) + CO_2(g).$$

(iv) At a given high temperature, which of the two carbonates, barium carbonate or calcium carbonate, would decompose more easily? Explain your answer.

Explanation:

Barium carbonate needs a higher temperature to decompose. This is because both Ba^{2+} and Ca^{2+} have the same charge but the cationic size of Ba^{2+} is bigger than that of Ca^{2+}. So, the Ca^{2+} ion has a greater charge density $(\propto \frac{q_+}{q_-})$, thus has greater polarizing power. It would distort the electron cloud of the CO_3^{2-} ion to a greater extent and weaken the intramolecular covalent bond more than what Ba^{2+} would do. Hence, a lower temperature is needed to decompose $CaCO_3$. In addition, the thermal decomposition is also driven by an increase in the entropy of the system at a higher temperature due to the formation of gas particles, CO_2.

When the Group 2 compounds decompose, they would form their respective metal oxides. Why is it so?

A: The metal oxide, MO, is a much more stable compound than the metal carbonate or nitrate or hydroxide. This is because the ionic bond in the MO is much stronger, due to the higher charge and smaller anionic size of the O^{2-} ion as compared to the larger size of CO_3^{2-} or the smaller charge and larger size of the NO_3^- and OH^- ions.

Do you know?

— Both $LiNO_3$ and Li_2CO_3 have a much lower decomposition temperature than their corresponding Group 1 nitrate or carbonate because of the high charge density of the Li^+ ion.

(v) How would you distinguish between solutions of barium chloride and calcium chloride? State in each case what you would see as a result of the test on each solution.

Explanation:

Add $H_2SO_4(aq)$ to each of the test tubes containing $BaCl_2(aq)$ and $CaCl_2(aq)$. The one that contains $BaCl_2(aq)$ will give a white ppt of $BaSO_4(s)$, whereas the one with $CaCl_2(aq)$ will not.

Do you know?

— The insoluble $CaSO_4(s)$ can be precipitated out if we use very concentrated $H_2SO_4(aq)$. This is because the K_{sp} value of $CaSO_4(s)$ is much larger than that of $BaSO_4(s)$, which thus needs a higher concentration of SO_4^{2-} for the ionic product of $CaSO_4(s)$ to surpass the K_{sp} value.

(b) 1.71 g of barium reacts with oxygen to form 2.11 g of an oxide **W**.

(i) Calculate the formula of **W**.

Explanation:

Amount of Ba in 1.71 g = 1.71/137 = 0.0125 mol.
Assuming the formula unit of oxide **W** = BaO_x.
Molar mass of oxide **W** = (137 + 16x) g mol^{-1}.
Amount of Ba in 2.11 g of oxide **W** = $\frac{2.11}{137+16x} = 0.0125\text{mol} \Rightarrow x = 2.0$.
Hence, the formula of oxide **W** is BaO_2 i.e., barium peroxide.

(ii) Give the formula of the anion present in **W**.

Explanation:

The anion present in **W** is O_2^{2-}.

(iii) What is the oxidation number of oxygen in this anion?

Explanation:

The oxidation number of oxygen in O_2^{2-} is −1.

(ii) Sodium forms an oxide, **Y**, which contains this same anion. Give the formula of **Y**.

Explanation:

The formula of **Y** is Na_2O_2.

(c) Treatment of either **W** or **Y** with dilute sulfuric acid leads to the formation of the sulfate of the metal, together with an aqueous solution of hydrogen peroxide, H_2O_2.

(i) Write an equation for the reaction of **Y** with dilute sulfuric acid.

Explanation:

$$Na_2O_2(s) + H_2SO_4(aq) \rightarrow Na_2SO_4(aq) + H_2O_2(aq).$$

Do you know?

— When $^{-}O{-}O^{-}$ reacts with H^+, the O^- part "captures" a H^+ to form an –OH group, just like an O^{2-} "captures" a H^+ to form OH^-.

(ii) The hydrogen peroxide solution produced may be separated from the other reaction product. Explain briefly why this is easier to achieve if **W** is used as the initial reagent rather than **Y**.

Explanation:

If **W** is used as the initial reagent rather than **Y**, the insoluble $BaSO_4(s)$ would form, which can be easily filtered out. This is unlike the $Na_2SO_4(aq)$, which is soluble in water.

(d) (i) Write an expression for K_p for the following equilibrium, giving the unit:

$$BaCO_3(s) \rightleftharpoons BaO(s) + CO_2(g).$$

Explanation:

$$BaCO_3(s) \rightleftharpoons BaO(s) + CO_2(g), \qquad K_p = P_{CO_2} \text{ atm}.$$

(ii) How would the numerical value of K_p change if $CaCO_3$ were used in place of $BaCO_3$ in *(d)(i)*? Explain your answer.

Explanation:

Since $CaCO_3$ requires a lower temperature to decompose than $BaCO_3$, at the same temperature, the partial pressure of CO_2 will be higher for $CaCO_3$. Hence, the K_p for $CaCO_3$ is going to be a larger value.

Do you know?

— The K_p for the decomposition of Group 2 carbonates correlate negatively with the thermal stability. This means that the more stable the carbonate, the smaller the K_p value. This also indicates that this relationship is going to hold for the Group 2 nitrates or hydroxides.

5. (a) (i) Define the term *lattice enthalpy*, illustrating your answer with reference to the oxide of a Group 2 metal.

Explanation:

Lattice enthalpy refers to the amount of energy that is released when one mole of ionic compound is formed from its constituent ions under standard conditions.

$$M^{2+}(g) + O^{2-}(g) \rightarrow MO(s).$$

Do you know?

— Enthalpy is equivalent to the term "heat."

(ii) What are the factors that affect the magnitude of lattice enthalpy?

Explanation:

The magnitude of lattice enthalpy depends on the both the cationic and anionic charges and both the cationic and anionic radii, i.e., L.E. $\propto \frac{q_+ \cdot q_-}{(r_+ + r_-)}$.

(iii) Why is the lattice enthalpy for a given compound found by use of the Born–Haber cycle rather than being measured by direct experiment?

Explanation:

The lattice enthalpy for a given compound is found by using the Born–Haber cycle rather than being measured by direct experiment because it is impossible to mix an isolated container of gaseous cations with another isolated container of anions and then measure the energy change.

So, is the lattice energy calculated from the Born–Haber cycle considered as an experimental or theoretical value?

A: It is considered as an experimental value because the other enthalpy terms that are used in the Born–Haber are experimental values.

(iv) Lattice enthalpies can be calculated from a formula based on a purely ionic model. Why do values calculated in this way often differ from those found from the Born–Haber cycle?

Explanation:

The theoretical lattice energy is calculated by assuming that the ionic compound has no covalent character. In reality, there is always a certain degree of covalent character in the ionic bond and this would make the experimental value different from the theoretical calculation.

Do you know?

— The theoretical model for the calculation of lattice energy also assumes that the ions are spherical in shape, separate entities, and each with its charge uniformly distributed, which in reality, they are not.

(b) (i) How does the solubility of the hydroxides of Group 2 metals change with increasing atomic number?

Explanation:

The solubility of the Group 2 hydroxides increases with increasing atomic number.

(ii) Suggest an explanation for this trend.

Explanation:

As we go down the group, the energy required to break up the ionic lattice of the Group 2 hydroxides decreases because of an increase in cationic radius. Hence, the ionic bond strength decreases. But the hydration enthalpy from the Group 2 cation also becomes less exothermic down the group, due to a decrease in charge density. In addition, as OH^- is a very small anion, the hydration enthalpy provided by this anion is very exothermic. Overall, as we go down the group, the hydration energies become much more exothermic than the rate of change of the lattice energy. As a result, the hydration enthalpies are more able to compensate for the energy required to break up the ionic lattice. Therefore, the solubility of the Group 2 hydroxides increases.

(c) (i) Explain why there is an increase in metallic character with an increase in atomic number in Group 14.

Explanation:

With an increase in atomic number in Group 14, the valence electrons become farther away from the nucleus. Hence, because these electrons are less strongly attracted, they are likely to delocalize. Therefore, there is an increase in metallic character.

Do you know?

— In fact, the metallic character increases for Groups 1, 2, 13, and 14 elements with an increase in atomic number down the group. Which means electropositivity increases, electronegativity decreases, inonization energy decreases, electron affinity decreases, reducing power increases, oxidizing power decreases, effective nuclear charge decreases, etc.

(ii) Explain why compounds of lead in oxidation state +4 are oxidizing.

Explanation:

Lead in oxidation state +4 has very high charge density; hence, it accepts electrons very readily, i.e., it undergoes reduction readily. This accounts for its strong oxidizing power.

Do you know?

— The strong oxidizing power of lead in the +4 oxidation state can be seen by their highly positive E^{θ} values:

$$Pb^{4+} + 2e^- \rightleftharpoons Pb^{2+}, \quad E^{\theta} = +1.69\ V, \quad \text{and}$$
$$PbO_2 + 4H^+ + 2e^- \rightleftharpoons Pb^{2+} + 2H_2O, \quad E^{\theta} = +1.47\ V.$$

— The compound with lead in the +4 oxidation state is a covalent compound and not an ionic one due to the high charge density of the Pb(IV) species. A solution containing Pb^{4+} ions is also highly acidic.

(d) From the position of radium in the periodic table, predict the following:

(i) the formula of radium carbonate;

Explanation:

Since radium comes from Group 2, the formula for radium carbonate is $RaCO_3$.

(ii) the equation for the thermal decomposition of solid radium carbonate; and

Explanation:

$$RaCO_3(s) \rightarrow RaO(s) + CO_2(g).$$

(iii) how the decomposition temperature required in *(d)(ii)* would compare with that required for magnesium carbonate.

Explanation:

The decomposition temperature for $RaCO_3$ is going to be higher than $MgCO_3$. This is because both Ra^{2+} and Mg^{2+} have the same charge but the cationic size of Ra^{2+} is bigger than Mg^{2+}. So, the Mg^{2+} ion has a greater charge density ($\propto \frac{q_+}{q_-}$), thus having greater polarizing power. It would distort the electron cloud of the CO_3^{2-} ion to a greater extent and weaken the intramolecular covalent bond more than what Ra^{2+} would do. Hence, a lower temperature is needed to decompose $MgCO_3$. In addition, the thermal decomposition is also driven by an increase in the entropy of the system at a higher temperature due to formation of gas particles, CO_2.

6. (a) Concentrated sulfuric acid reacts with sodium chloride as follows:

$$H_2SO_4 + Cl^- \rightleftharpoons HCl + HSO_4^-.$$

(i) Identify the conjugate acid–base pairs in this reaction.

Explanation:

The conjugate acid–base pairs are: $HCl/\ Cl^-$ and $H_2SO_4/\ HSO_4^-$.

Do you know?

— A conjugate acid–base pair only differs by a H^+.

(ii) What would be the observable result of this reaction?

Explanation:

White fumes of HCl gas would be observed.

Do you know?

— Concentrated sulfuric acid has a higher boiling point than liquid HCl because the hydrogen bonds between the sulfuric acid molecules are stronger than the permanent dipole–permanent dipole interaction between the polar HCl molecules. Hence, concentrated sulfuric acid is less volatile than HCl. The above method is a common method to produce HCl gas.

Q So, is the above reaction an acid–base reaction?

A: Yes, of course. H_2SO_4 is the acid, according to Brϕnsted–Lowry definition, while Cl^- is the base. Similarly, in the backward reaction, HCl is the acid while HSO_4^- is the Brϕnsted–Lowry base. So in effect, an acid reacts with a base to give another pair of acid–base!

(iii) Explain why this reaction goes almost completely to the right despite the fact that both hydrochloric and sulfuric acids are strong.

Explanation:

As the HCl is more volatile than the H_2SO_4, the vaporization of the HCl would drive the position of equilibrium, $H_2SO_4 + Cl^- \rightleftharpoons HCl + HSO_4^-$, toward the right according to Le Chatelier's Principle.

(b) When concentrated sulfuric acid reacts with solid sodium iodide, hydrogen iodide, hydrogen sulfide, sodium hydrogensulfate, and water are formed together with one other product.

(i) Identify this product and state how you would recognize it.

Explanation:

The product is solid iodine and it is black in color. Or, it can evolve as violet vapor.

(ii) Write an ionic half-equation to show the conversion of sulfuric acid into hydrogen sulfide.

Explanation:

$$H_2SO_4 + 8H^+ + 8e^- \rightarrow H_2S + 4H_2O.$$

(iii) Hence, write the full ionic equation for the reaction between concentrated sulfuric acid and sodium iodide.

Explanation:

$$9I^- + 10H_2SO_4 \rightarrow 9HSO_4^- + HI + H_2S + 4I_2 + 4H_2O.$$

Do you know?

— The oxidizing power of the Group 17 halogens decreases down the group, which means that the reducing power of the Group 17 halogens would increase down the group. This can be observed in the greater likelihood of the element to form IO_3^- than ClO_3^- at the same temperature.

— Similarly, if a halogen is less likely to undergo reduction, then the corresponding halide would be more likely to undergo oxidation. This is easy to understand as reduction means taking in electrons. If the atom is less likely to take in electrons, it can only mean that the extra electron is less strongly attracted, which would also mean that it is more likely to remove an electron from this same atom. Thus, the weaker the oxidizing agent, the less likely it will be reduced. Then, the reduced form of this oxidizing agent will be a stronger reducing agent!

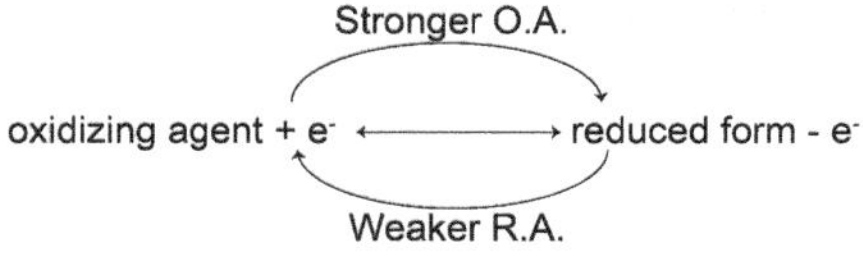

(iv) What is the function of sulfuric acid in this reaction?

Explanation:

Since I^- is oxidized to form I_2, the sulfuric acid is an oxidizing agent as it is reduced to H_2S itself.

(c) Explain concisely why the type of reaction occurring in *(b)* does not occur with sodium chloride.

Explanation:

The type of reaction occurring in *(b)* does not occur with sodium chloride because Cl^- is a weaker reducing agent than I^- as the valence electrons of Cl^- are more strongly attracted; hence, they are more difficult to be removed during the oxidation of Cl^-.

CHAPTER 11

INTRODUCTION TO TRANSITION METALS AND THEIR CHEMISTRY

What do you know regarding the transition metals and their chemistry?

(1) Some special features of transition metals

- A transition element is defined as an element that possesses a partially filled *d* subshell in at least one of its stable species. Scandium is not a transition metal as its only common ion, Sc^{3+}, does not have any *d* electrons. Zinc is also not a transition metal as it contains a fully filled 3*d* subshell in its only oxidation state of +2.
- Electronic configuration of the transition metals:

Element	(Sc)	Ti	V	Cr	Mn
Electronic configuration	$[Ar]3d^14s^2$	$[Ar]3d^24s^2$	$[Ar]3d^34s^2$	$[Ar]3d^54s^1$	$[Ar]3d^54s^2$
Element	Fe	Co	Ni	Cu	(Zn)
Electronic configuration	$[Ar]3d^64s^2$	$[Ar]3d^74s^2$	$[Ar]3d^84s^2$	$[Ar]3d^{10}4s^1$	$[Ar]3d^{10}4s^2$

The electronic configurations of Cr and Cu have only one 4*s* electron as it is more energetically favorable for them to have a half-filled or fully filled 3*d* subshell, respectively. This is because a half-filled or fully filled 3*d* subshell has a symmetrical distribution of electron density around the nucleus, which has a higher stability or lower energy level.

(2) Physical properties of transition metals

(i) *Atomic radius*

- The transition metal has an atomic radius which is much smaller than the *s* block elements. This is because the shielding effect on the valence electrons of *s* block elements comes from the *s* and *p* electrons, whereas the transition metals experience additional shielding effect from the *d* electrons, which are actually poorer shielders when compared to the *s* and *p* electrons. This is because a *d* orbital is more diffuse. Hence, the effective nuclear charge (ENC) experienced by the valence electrons is much stronger for the transition metals.

— But as we traverse across the *d* block, the atomic radii are almost constant. This is because other than the increase of the nuclear charge across the *d* block, the shielding effect also increases as electrons are being added to the penultimate (the one before the last) subshell. This increase in shielding effect offset the increase in nuclear charge, resulting in a relatively constant ENC experienced by the valence electrons.

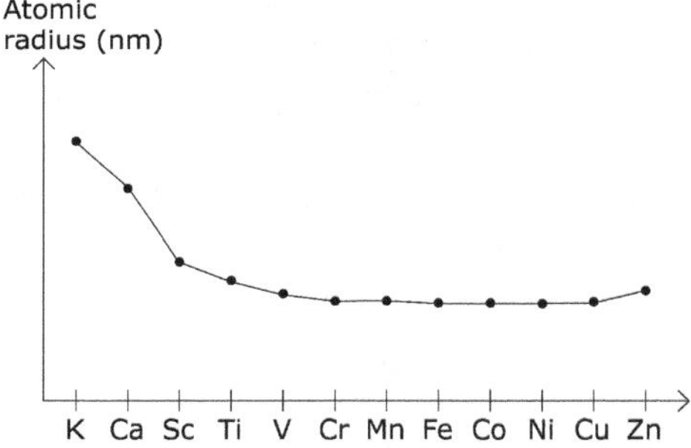

(ii) *First ionization energy*

The first ionization energy is nearly constant for the transition metals. The explanation is the same as the constant trend in atomic radii.

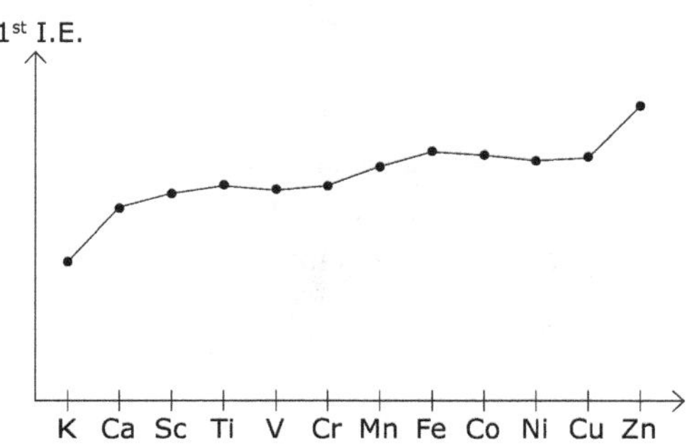

The slight irregularities in the first ionization energy can be attributed to: (1) ENC effect; (2) inter-electronic repulsion between an electron pair residing in the *d* orbital; or (3) a 4*s* electron having a higher energy than a 3*d* electron.

(iii) *Melting point*

The melting point of transition metals is much higher than that of the *s* block metals because of the smaller atomic radii and greater number of valence electrons that are available for delocalization in the metallic bonding.

(iv) *Electrical conductivity*

Transition metals are good electrical conductors of electricity because of the greater number of delocalized electrons which are available as charge carriers.

(3) Chemical properties of transition metals

(i) *Variable oxidation states*

Due to the close proximity of the 3*d* and 4*s* subshells, it is energetically feasible for transition metals to lose or use a different number of valence electrons for bonding. As a result, these elements are able to exist in variable oxidation states. For example, Fe^{2+} and Fe^{3+}; and Mn^{2+} and MnO_4^{2-}.

(ii) *Catalytic property*

— Heterogeneous catalysis

In heterogeneous catalysis, the transition elements possess a partially filled *d* subshell. This allows the reactants to form weak bonds by donating electron density into the partially filled *d* subshell when they adsorb on the active sites. Because of the weak bonds that are formed, the intramolecular bonds within the reactant molecule itself are weakened. Hence, this lowers the amount of energy that is required to break the bonds. In addition, the catalyst also lowers the activation energy by:

- orientating the reactant particles so that they achieve the correct collision geometry; and
- increasing the concentrations of the reactant particles locally.

Two good examples to quote are, the Haber process which uses Fe catalyst and the hydrogenation of alkenes which uses Ni catalyst.

— Homogeneous catalysis

In homogeneous catalysis, the transition metals exhibit variable oxidation states. As such, the element is able to accept or lose electrons with ease when transiting between these different oxidation states. Such an alternative mechanism has a lowered activation barrier as compared to the uncatalyzed reaction. An example is the participation of Fe^{2+}or Fe^{3+}in the reaction of I^- and $S_2O_8^{2-}$.

(iii) *Complex formation*

— Due to the availability of the partially filled *d* subshell, transition elements are able to form multiple complexes by accepting a lone pair of electrons from the ligand molecules. Dative covalent bonds bind the transition metal and the ligand together.

— Ligands can be monodendate, bidendate, polydendate, etc. A mondendate ligand has only one atom that provides a lone pair of electrons for dative covalent bond formation, a bidendate has two atoms, and so on and so forth.

— Octahedral and tetrahedral complexes refer to complexes containing six and four dative covalent bonds, respectively, being formed. The number of dative covalent bonds is also known as the <u>coordination number</u>. The following examples show the formation of two octahedral complexes with three bidentate ligands but with a coordination number of six:

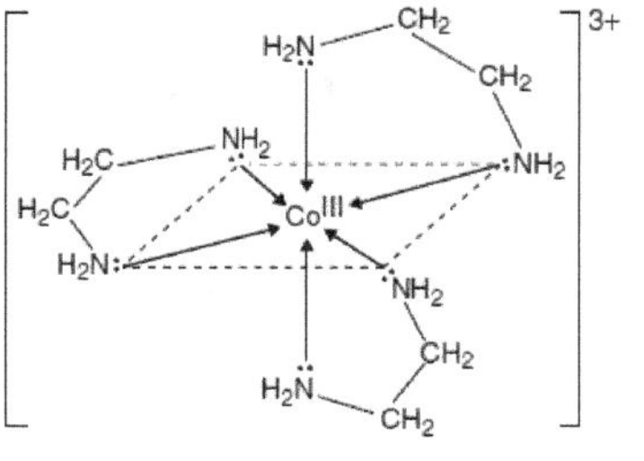

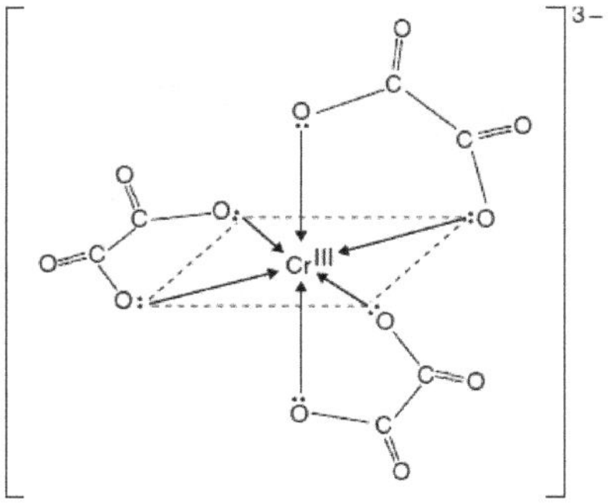

— The types and shapes of complexes formed depend on: (1) the type of ligand; (2) the nature of the transition element; and

(3) the charge of the transition element. The overall charge of a complex would depend on the charge of the metal center and the total charges of the ligands. Examples of some common complexes are:

Coordination number	Examples	Shape
6	$[Fe(CN)_6]^{3-}$, $[Cu(EDTA)]^{2-}$	Octahedral
4	$[CuCl_4]^{2-}$, $[Cu(NH_3)_4]^{2+}$	Square planar or tetrahedral (depending on reaction conditions)
2	$[Ag(NH_3)_2]^{+}$, $[CuCl_2]^{-}$	Linear

— If an incoming ligand can form a more stable complex than any of the existing ligands, it would be able to displace that existing ligand. Such a process is known as ligand exchange. The stability of a complex is determined by the equilibrium constant known as $K_{stability}$, which is a measurement of the strength of interaction between the ligands and metal center. The greater the stability constant, the higher the concentration of the complex, and the more stable the complex that is formed. An example is the displacement of the O_2 molecule of oxyhemoglobin by the poisonous CO molecule to form carboxyhemoglobin, as the CO molecule is a stronger ligand than O_2.

— Different complexes that are formed from the same transition metal center but with different ligands have different abilities to undergo reduction. Examples are:

$$[Fe(H_2O)_6]^{3+}(aq) + e^- \rightleftharpoons [Fe(H_2O)_6]^{2+}(aq), \quad E^{\theta} = +0.77\text{ V, and}$$
$$[Fe(CN)_6]^{3-}(aq) + e^- \rightleftharpoons [Fe(CN)_6]^{4-}(aq), \quad E^{\theta} = +0.36\text{ V.}$$

(iv) *Color of complexes*

Due to the presence of a partially filled *d* subshell, under the influenced of an asymmetrical electrostatic field (crystal field theory), the five degenerate *d* orbitals become non-degenerate and separate into two different energy levels, with the energy gap, $\Delta E = \frac{hc}{\lambda}$:

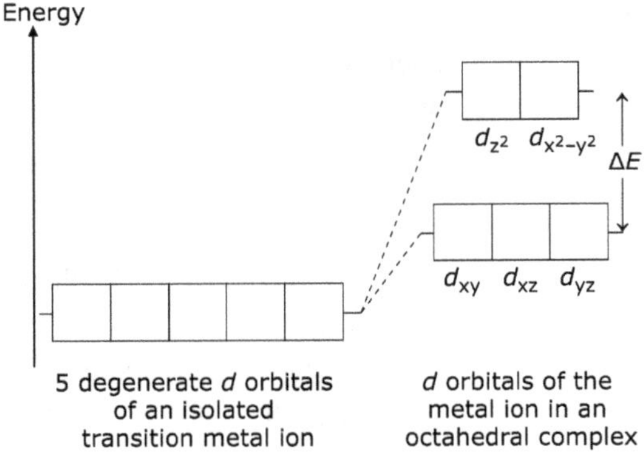

Electrons in the lower energy level can absorb a photon of energy, with the wavelength within the visible range of the electromagnetic spectrum, and are promoted to the higher energy level. The color of the compound observed is a complementary color to the wavelength of the radiation being absorbed. Such an electronic transition is known as *d–d* transition. Examples of some colored complexes are:

$[Cu(H_2O)_6]^{2+}$: light blue; $[CuCl_4]^{2-}$: yellow; $[Cu(NH_3)_4]^{2+}$: dark blue; $[Fe(H_2O)_6]^{2+}$: green; $[Fe(H_2O)_6]^{3+}$: yellow; and $[Fe(SCN)(H_2O)_5]^{2+}$: blood red.

The longer the wavelength of the observed color, the shorter the wavelength of the complementary radiation being absorbed. This would mean that the energy gap of the splitting of the *d* subshell that is being created by the ligands becomes larger. The gap depends on the following factors:

- nature of the transition metal ion;
- oxidation state of the metal ion, e.g., $[Fe(H_2O)_6]^{3+}$ is yellow and $[Fe(H_2O)_6]^{2+}$ is pale green;
- orientation of the ligands around the central metal ion, i.e., type of geometry; and
- type of ligands.

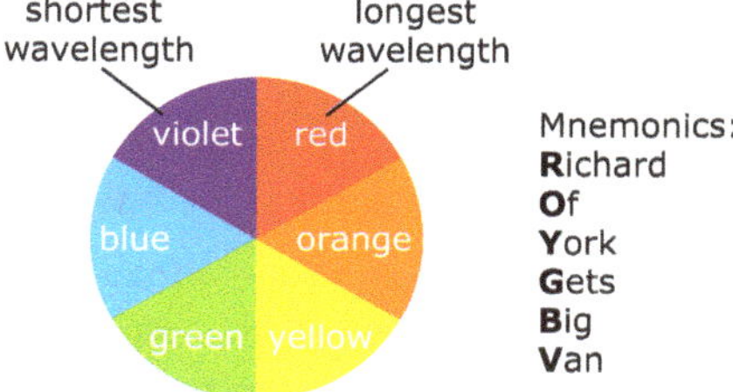

Transition elements with compounds having the d^0 and d^{10} configurations would appear colorless as *d–d* transition cannot take place.

1. (a) Discuss, in relation to their electronic structures, the variation in the number of oxidation states shown by the elements along the first transition series.

Explanation:

Due to the close proximity of the 3*d* and 4*s* subshells, it is energetically feasible for transition metals to lose or use a different number of valence electrons for bonding. As a result, these elements are able to exist in variable oxidation states. For example, Fe^{2+} and Fe^{3+}; and Mn^{2+} and MnO_4^{2-}.

Do you know?

— The maximum number of oxidation states of a transition element, from Sc to Mn, correspond to the total number of 3*d* and 4*s* electrons the atom has. For example, Mn has a total of seven valence electrons and its maximum oxidation state is +7 as in the MnO_4^- ion. If the oxidation state of the element is more than +3, it is likely that the transition element utilizes its valence electrons to form covalent bonds. It is very energetically demanding to lose more than three electrons to form a cation with a +4 charge!

— Transition metal cations with a +2 charge are weakly acidic because of their high charge density, which arises from the small cationic radius. But a +3 cation undergoes appreciable hydrolysis to give a relatively acidic solution that is strong enough to decompose carbonate or hydrogencarbonate.

Why does the maximum oxidation state of transition elements after Mn not correspond to the maximum number of valence electrons that they have?

A: Usually, an oxidation state that is more than +3 involves the atom undergoing covalent bond formation with other atoms. If you want Fe atom, which has a total of eight valence electrons, to assume the maximum oxidation state of +8, then the Fe atom needs to take in energy and promote some of the electrons into subshells with a higher energy level. No doubt after bonds are formed, energy would be released. But is the amount of "energy return" viable to compensate for the "energy investment?" In addition, do not forget that using eight electrons to form eight covalent bonds would result in a total of 16 electrons on top of the eight electrons that are already present in the 3*s* and 3*p* subshells. Can you imagine how much inter-electronic repulsion that is present among the 24 electrons that are crowding around the $n = 3$ principal quantum shell? So, according to the "investment-return theory," it is energetically non-feasible for the elements after Mn to assume their maximum oxidation state that corresponds to the number of valence electrons that they have under normal conditions.

(b) (i) What is the oxidation number of nickel in each of the following compounds?

(A) $K_2Ni(CN)_4$,

(B) $Ni(H_2O)_6(NO_3)_2$, and

(C) $Ni(CO)_4$.

Explanation:

To calculate the oxidation states, you need to know the charge of each of the ligands and the overall charge of the complex. The complex in (A) is $[Ni(CN)_4]^{2-}$, (B) is $[Ni(H_2O)_6]^{2+}$, and (C) is $Ni(CO)_4$.

Hence, the oxidation number of Ni in (A) is +2 (since the ligand is CN^-); in (B) is +2 (since H_2O is a neutral molecule), and in (C) is 0 (since CO is a neutral molecule).

Do you know?

— The coordination number of the complex in $[Ni(CN)_4]^{2-}$ is four; in $[Ni(H_2O)_6]^{2+}$ is six, and in Ni(CO) is four. The coordination numbers of the complexes here correspond to the number of ligands that each of the complex has, as all these ligands are monodentate.

(ii) Suggest the likely stereochemical arrangements of ligands around the nickel atoms in compounds (A) and (B).

Explanation:

The stereochemical arrangement of ligands around the nickel atom in compound (A) is tetrahedral while it is octahedral in compound (B).

(iii) Nickel carbonyl, (C), has a tetrahedral arrangement of the CO ligands around the nickel atom. What would you expect the stereochemical arrangement around the nickel atom to be in the compound $K_4Ni(CN)_4$? Explain your answer.

Explanation:

The stereochemical arrangement around the nickel atom in the compound $K_4Ni(CN)_4$ should also be tetrahedral as there are four bonds around the nickel atom, similar to that in $Ni(CO)_4$.

(c) A green aqueous solution of a nickel(II) salt is converted to a blue solution, containing $[Ni(NH_3)_6]^{2+}$ ions by the addition of an excess of aqueous ammonia. The green solution is then converted to a yellow solution, containing $[Ni(CN)_4]^{2-}$ ions by addition of an excess of aqueous potassium cyanide. Explain why the colors of the solutions are different.

Explanation:

The green complex in aqueous solution corresponds to $[Ni(H_2O)_6]^{2+}$. The following shows the colors of visible radiation being absorbed by the respective complex:

	Color observed	Color absorbed
$[Ni(H_2O)_6]^{2+}$	Green	Red
$[Ni(NH_3)_6]^{2+}$	Blue	Orange
$[Ni(CN)_4]^{2-}$	Yellow	Violet

The energy gap of the complexes caused by the ligands decreases from $[Ni(CN)_4]^{2-} > [Ni(NH_3)_6]^{2+} > [Ni(H_2O)_6]^{2+}$. This trend shows that the splitting of the *d* subshell, as brought about by the ligands, decreases in the order: $CN^- > NH_3 > H_2O$.

Why did the $[Ni(NH_3)_6]^{2+}$ formed in place of the $[Ni(H_2O)_6]^{2+}$ when aqueous ammonia is added?

A: The phenomenon in which the ligand of a complex is being displaced by another stronger ligand is known as ligand exchange. In aqueous ammonia, we have the following equilibrium:

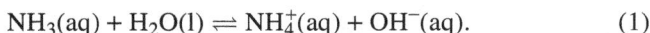

$$NH_3(aq) + H_2O(l) \rightleftharpoons NH_4^+(aq) + OH^-(aq). \quad (1)$$

When $NH_3(aq)$ is added to the solution containing the $[Ni(H_2O)_6]^{2+}$ complex, a green ppt of $Ni(OH)_2$ is formed first when the ionic product exceeds its K_{sp}:

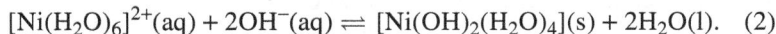

$$[Ni(H_2O)_6]^{2+}(aq) + 2OH^-(aq) \rightleftharpoons [Ni(OH)_2(H_2O)_4](s) + 2H_2O(l). \quad (2)$$

Then, when excess $NH_3(aq)$ is added, the following ligand exchange takes place:

$$[Ni(H_2O)_6]^{2+}(aq) + 6NH_3(aq) \rightleftharpoons [Ni(NH_3)_6]^{2+}(aq) + 6H_2O(l). \quad (3)$$

As a result, the $[Ni(H_2O)_6]^{2+}$ decreases and according to Le Chatelier's Principle, the position of the equilibrium for reaction (2) shifts to the left, causing the $Ni(OH)_2$ to dissolve. Hence, a blue solution consisting of $[Ni(NH_3)_6]^{2+}(aq)$ is formed.

Now, when an excess of KCN(aq) is added, the following ligand exchange takes place:

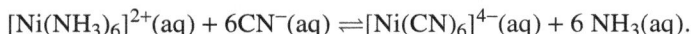

$$[Ni(NH_3)_6]^{2+}(aq) + 6CN^-(aq) \rightleftharpoons [Ni(CN)_6]^{4-}(aq) + 6\ NH_3(aq).$$

Hence, the fact that NH_3 is able to displace the H_2O ligand while the CN^- is able to displace the NH_3 ligand, is a demonstration of the decreasing ligand strength in the order: $CN^- > NH_3 > H_2O$. This also means that the complex stability constant, $K_{\text{stability}}$, also decreases in the order: $[Ni(CN)_6]^{4-}(aq) > [Ni(NH_3)_6]^{2+}(aq) > [Ni(H_2O)_6]^{2+}(aq)$.

Q So, the precipitation of $Ni(OH)_2$ can be considered as a ligand exchange reaction? Can we use this explanation to account for the question pertaining to ionic equilibria, which involves the K_{sp}?

A: Yes, you can perceive precipitation as a kind of ligand exchange! But when writing the solubility equation in ionic equilibria, for simplicity, just ignore the ligand exchange part. Just treat it as, $Ni^{2+}(aq) + 2OH^-(aq) \rightleftharpoons Ni(OH)_2(s)$. And remember that percipitation occurs when ionic product exceeds K_{sp} value!

Q Based on the above, can we say that the OH^- is a stronger ligand than H_2O?

A: Yes, you can say that. This is because the OH^- is a negatively charged species; hence, it is more electron-rich than the neutral H_2O. Therefore, the OH^- is more likely to form a dative covalent bond with the metal center than the H_2O.

Q What happen if $H_2SO_4(aq)$ is now slowly added to a solution containing the blue $[Ni(NH_3)_6]^{2+}(aq)$ complex in excess aqueous ammonia?

A: Before $H_2SO_4(aq)$ is added, we have the following major equilibria existing in the system:

$$NH_3(aq) + H_2O(l) \rightleftharpoons NH_4^+(aq) + OH^-(aq), \text{ and} \tag{1}$$

$$[Ni(H_2O)_6]^{2+}(aq) + 6NH_3(aq) \rightleftharpoons [Ni(NH_3)_6]^{2+}(aq) + 6H_2O(l). \tag{2}$$

When $H_3O^+(aq)$ is added slowly, the following acid–base reaction takes place:

$$H_3O^+(aq) + NH_3(aq) \rightarrow NH_4^+(aq) + H_2O(l).$$

As a result, [NH_3(aq)] decreases. According to Le Chatelier's Principle, the position of the equilibrium for reaction (2) shifts to the left. The blue coloration disappears and we have $[Ni(H_2O)_6]^{2+}$(aq) accumulating. Thus, the solution now becomes green in color because of the $[Ni(H_2O)_6]^{2+}$(aq) complex that is formed. But with higher $[Ni(H_2O)_6]^{2+}$(aq), the ionic product of the following equilibria exceeds its K_{sp} value. Hence, a green ppt of $Ni(OH)_2$ is formed:

$$[Ni(H_2O)_6]^{2+}(aq) + 2OH^-(aq) \rightleftharpoons [Ni(OH)_2(H_2O)_4](s) + 2H_2O(l). \quad (3)$$

As more H_3O^+(aq) is added slowly, the following acid–base reaction takes place:

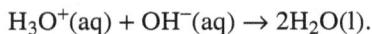

$$H_3O^+(aq) + OH^-(aq) \rightarrow 2H_2O(l).$$

As a result, [OH^-(aq)] decreases. According to Le Chatelier's Principle, the position of the equilibrium for reaction (3) shifts to the left. This causes the green ppt of $Ni(OH)_2$ to dissolve. Hence, we get the green solution consisting of the $[Ni(H_2O)_6]^{2+}$(aq) complex.

Do you know?

— The amount of splitting caused by the various ligands that are bonded to a given metal center is presented in the spectrochemical series that is partially listed here:

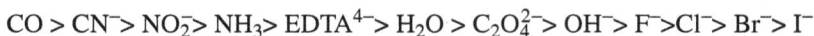

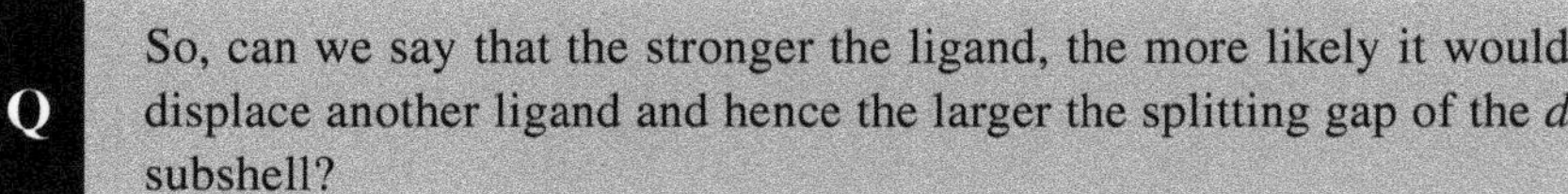

stronger field ligands that cause large splitting (large ΔE) — weaker field ligands that cause small splitting (small ΔE)

Q So, can we say that the stronger the ligand, the more likely it would displace another ligand and hence the larger the splitting gap of the *d* subshell?

A: No, the correlation is not necessary 100%. There can be a ligand that causes a large energy gap in the splitting of the *d* subshell than another ligand. But the latter ligand is actually a stronger ligand that can displace the former.

This is because the splitting of the *d* subshell depends on the following variables:

- nature of the transition metal ion;
- oxidation state of the metal ion, e.g., $[Fe(H_2O)_6]^{3+}$ is yellow and $[Fe(H_2O)_6]^{2+}$ is pale green;
- orientation of the ligands around the central metal ion, i.e., type of geometry; and
- type of ligands.

2. In addition to its widespread use as a structural material, iron is often employed as a heterogeneous catalyst for industrial reactions.
   (a) What is meant by the term *heterogeneous catalysis*?

Explanation:

In heterogeneous catalysis, the catalyst and the reactants are in different phases.

(b) Explain how iron may function as a heterogeneous catalyst in the Haber process for the synthesis of ammonia.

Explanation:

In the Haber process, the iron element possesses a partially filled *d* subshell. This allows the N_2 and H_2 molecules to form weak bonds with the catalyst surface when they adsorb on the active sites. Because of the weak bonds that are formed, the intramolecular bonds within the reactant molecule itself are weakened. Hence, this lowers the amount of energy that is required to break the bonds. In addition, the catalyst lowers the activation energy by:

- orientating reactant particles so that they achieve the correct collision geometry; and
- increasing the concentrations of the reactant particles locally.

Do you know?

— The activation energy of the formation of ammonia from nitrogen and hydrogen is high because of the requirement to break the strong N≡N triple bond.

(c) What chemical properties of iron make it effective both as a heterogeneous and a homogeneous catalyst?

Explanation:

The partially filled *d* subshell enables iron to assume variable oxidation states when acting as a homogeneous catalyst. In addition, the partially filled *d* subshell also enables iron to accept electron pairs from reactants adsorbing on its active sites when it functions as a heterogeneous catalyst.

Do you know?

— It is because of the "partially filled *d* subshell," that the transition element is able to:

(1) exhibit variable oxidation states;
(2) act as catalyst;
(3) form various complexes; and
(4) give different colored compounds through *d–d* transition.

(d) Explain, by means of one example, why iron complexes are essential to human life.

Explanation:

The hemoglobin that is responsible for the transport of oxygen molecules in the red blood cells consists of an iron(II) ion coordinated to six groups of molecules. Four of the coordination sites are taken up by the nitrogen atoms from a ring system called a porphyrin, which acts as a tetradentate ligand. This complex is called 'haem.'

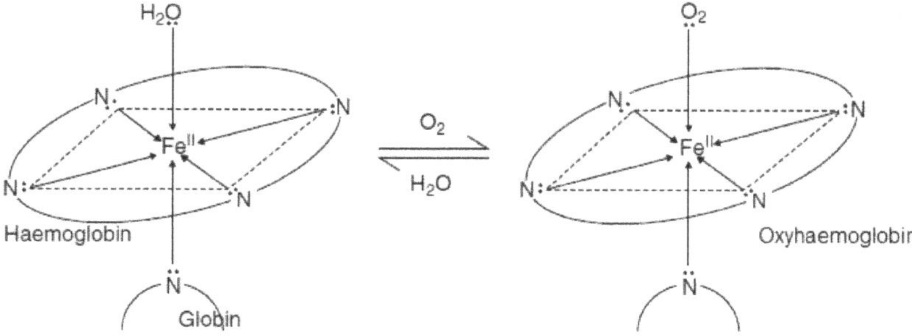

The fifth coordination site is taken up by the nitrogen atom of a complex protein called 'globin.' The sixth co-ordination site is reversibly bonded to an oxygen molecule. This thus allows the hemoglobin to carry oxygen from one part of the body to another.

Do you know?

— The better ligands such as carbon monoxide and cyanide ions can be bonded less reversibly to the metal center, which thus inhibit the hemoglobin from carrying oxygen. Hence, this accounts for the toxic nature of such strong ligands.

— These ligands are powerful ligands because the covalent bond that is formed between the ligand and the metal center is very strong due to the outflow of electron density from the metal center back to the ligand through a phenomenon we called "back-bonding."

Q So, when we have CO or CN^- poisoning, does that mean there is no cure at all?

A: No. When you suffer from CO poisoning, all you need to do is to breathe in pure oxygen. Under a high partial pressure of O_2, the position of the equilibrium for the formation of carboxyhemoglobin would shift left toward the oxyhemoglobin:

$$O_2\text{–Hb} + CO \rightleftharpoons CO\text{–Hb} + O_2.$$

Similarly, for CN^- poisoning, $EDTA^{4-}$ solution is usually being administered. The hexadentate $EDTA^{4-}$ is a more powerful ligand than the monodentate CN^- because once it is bonded to the metal center, in order to remove it, you need to break the six covalent bonds at one ago. This is statistically not very feasible.

Q But wouldn't the $EDTA^{4-}$ solution also remove the Ca^{2+} in our body?

A: Yes, you are right about this. That is why we cannot administer too much of the solution at one go.

3. Replacement of water molecules by other ligands generally changes the redox potentials of transition metal ions. Use the following information in your answer to this question.

	E^θ/V
$[Co(H_2O)_6]^{3+}(aq) + e^- \rightleftharpoons [Co(H_2O)_6]^{2+}(aq)$,	+1.81
$1/2O_2(g) + 2H^+(aq) + 2e^- \rightleftharpoons H_2O(l)$,	+1.23
$H^+(aq) + e^- \rightleftharpoons 1/2H_2(g)$,	0.00
$[Co(CN)_6]^{3-}(aq) + e^- \rightleftharpoons [Co(CN)_6]^{4-}(aq)$,	-0.83

(a) What products are likely to be formed when cobalt(III) sulfate is dissolved in water? Give an equation for the reaction.

Explanation:

Based on the given E^θ values, $[Co(H_2O)_6]^{3+}(aq)$ would undergo reduction while $H_2O(l)$ would undergo oxidation.

Reduction half-equation: $[Co(H_2O)_6]^{3+}(aq) + e^- \rightarrow [Co(H_2O)_6]^{2+}(aq)$.
Oxidation half-equation: $H_2O(l) \rightarrow 1/2O_2(g) + 2H^+(aq) + 2e^-$.

Overall redox equation:

$$2[Co(H_2O)_6]^{3+}(aq) + H_2O(l) \rightarrow 2[Co(H_2O)_6]^{2+}(aq) + 1/2O_2(g) + 2H^+(aq).$$

A pink $[Co(H_2O)_6]^{2+}$(aq) complex and O_2 gas would be formed.

Do you know?

— Calculating the E^{θ}_{cell} value for the above reaction:

$$E^{\theta}_{cell} = E^{\theta}_{Red} - E^{\theta}_{Ox} = 1.81 - (+1.23) = +0.58 \text{ V}.$$

Since $E^{\theta}_{cell} > 0$, the reaction is thermodynamically spontaneous under standard conditions.

Q Why is the E^{θ} value for $[Co(H_2O)_6]^{3+}$(aq) more positive than that for $[Co(CN)_6]^{3-}$(aq)?

A: A more positive E^{θ} value means that the species is more likely to undergo reduction. Since $[Co(H_2O)_6]^{3+}$(aq) is a positively charged complex while $[Co(CN)_6]^{3-}$(aq) is negatively charged, it is easier for a positively charged complex to take in electrons than a negatively charged one! Similarly, the standard reduction potential for Fe^{3+} is more positive than $Fe(OH)_3$ is because it is easier for a positive species to take in an electron than a neutral one:

$$[Fe(H_2O)_6]^{3+} + e^- \rightleftharpoons [Fe(H_2O)_6]^{2+}, \quad E^{\theta} = +0.77 \text{ V, and}$$
$$Fe(OH)_3 + e^- \rightleftharpoons Fe(OH)_2 + OH^-, \quad E^{\theta} = -0.56 \text{ V.}$$

Q How do we know whether a particular oxidation state of the complex is stable in water?

A: You can compare the standard reduction potential of the complex against that for H_2O, i.e. $1/2O_2(g) + 2H^+(aq) + 2e^- \rightleftharpoons H_2O(l)$ $E^{\theta} = +1.23$V. If the standard reduction potential is more positive than +1.23 V, then the higher oxidation state of the complex would likely undergo reduction in the presence of water to form a complex of lower oxidation state. If it is less positive than +1.23 V, it would be oxidized by O_2 in the air. Similarly, if the standard reduction potential is less positive than the value for acid, i.e., $H^+(aq) + e^- \rightleftharpoons 1/2H_2(g)$ $E^{\theta} =$ 0.00 V, then the lower oxidation state of the complex species would react under acidic conditions to form a higher-oxidation-state species. Hence, the lower-oxidation-state complex becomes less stable in the presence of acid. Or from another perspective, we can say that acidic conditions favor or stabilize the higher oxidation state of the complex.

(b) If an aqueous solution of cobalt(II) chloride is mixed with an excess of aqueous potassium cyanide, the ion $[Co(CN)_6]^{4-}(aq)$ is formed. Explain why this mixture absorbs oxygen from the air.

Explanation:

Comparing the standard reduction potential:

	E^θ/V
$1/2O_2(g) + 2H^+(aq) + 2e^- \rightleftharpoons H_2O(l)$,	+1.23, and
$[Co(CN)_6]^{3-}(aq) + e^- \rightleftharpoons [Co(CN)_6]^{4-}(aq)$,	–0.83.

$O_2(g)$ will undergo reduction while $[Co(CN)_6]^{4-}(aq)$ will undergo oxidation:

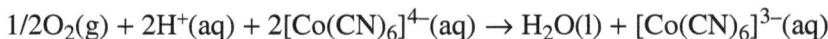

$$1/2O_2(g) + 2H^+(aq) + 2[Co(CN)_6]^{4-}(aq) \rightarrow H_2O(l) + [Co(CN)_6]^{3-}(aq)$$

The $E^\theta_{cell} = E^\theta_{Red} - E^\theta_{Ox} = 1.23 - (-0.83) = +2.06$ V.

Since $E^\theta_{cell} > 0$, the reaction is thermodynamically spontaneous under standard conditions, the mixture absorbs oxygen from the air.

Do you know?

— There are lots of cosmetics that contain transition metal ions and to prevent the cosmetics from deteriorating in air, they are commonly converted to complexes that are less likely to react with air. For example:

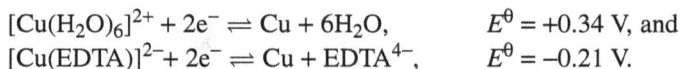

$[Cu(H_2O)_6]^{2+} + 2e^- \rightleftharpoons Cu + 6H_2O$,	$E^\theta = +0.34$ V, and
$[Cu(EDTA)]^{2-} + 2e^- \rightleftharpoons Cu + EDTA^{4-}$,	$E^\theta = -0.21$ V.

Since the standard reduction potential of $[Cu(EDTA)]^{2-}$ is less positive than that for $[Cu(H_2O)_6]^{2+}$, the Cu^{2+} is stabilized.

(c) Suggest another reaction which might take place in the cyanide mixture if air were excluded.

Explanation:

$[Co(CN)_6]^{3-}(aq) + e^- \rightleftharpoons [Co(CN)_6]^{4-}(aq)$, $E^\theta = -0.83$ V, and
$2H_2O + 2e^- \rightleftharpoons H_2 + 2OH^-$, $E^\theta = -0.83$ V.

If air is excluded, then the $[Co(CN)_6]^{4-}(aq)$ may reduce the H_2O to H_2:

$$2H_2O + 2[Co(CN)_6]^{4-} \rightarrow H_2 + 2OH^- + 2[Co(CN)_6]^{3-}.$$

The $E^\theta_{cell} = E^\theta_{Red} - E^\theta_{Ox} = -0.83 - (-0.83) = +0.0$ V.
Since $E^\theta_{cell} = 0$, the reaction would reach an equilibrium state.

Q Why not consider the reduction of H^+ by the $[Co(CN)_6]^{4-}(aq)$ instead?

A: The concentration of H^+ is very low in the mixture, hence the reaction is unlikely to take place. But if an acid is added, then the reaction becomes thermodynamically spontaneous as shown by the following E^θ_{cell} calculation:

$$E^\theta_{cell} = E^\theta_{Red} - E^\theta_{Ox} = 0.00 - (-0.83) = +0.83 \text{ V}.$$

Q But since the calculated $E^\theta_{cell} = 0$, shouldn't the reaction not likely to proceed?

A: Well, the reaction is not thermodynamically spontaneous under standard conditions. But in view of the large amount of water molecules that is present, the reduction potential value for $2H_2O + 2e^- \rightleftharpoons H_2 + 2OH^-$ may become more positive than –0.83 V, hence making the $E_{cell} > 0$. Thus, the reaction may still proceed under non-standard conditions.

4. One difference between the chemistries of calcium and manganese is that manganese can undergo disproportionation reactions whereas calcium cannot.

   (a) Explain in terms of the electronic structures of calcium and manganese why disproportionation is a feature of the chemistry of many of the elements of the *d* block but none of those in the *s* block.

Explanation:

Ca has only two valence electrons, thus it is viable for it to lose these two electrons in order to achieve the stable octet configuration, with an oxidation state of +2 only. On the contrary, Mn's electronic configuration is $[Ar]3d^54s^2$, and due to the proximity of the energy levels of the 3*d* and 4*s* subshells, it is possible for Mn to exhibit variable oxidation states. Hence, because of these different oxidation states, it is possible for transition elements of intermediate oxidation states to undergo disproportionation to form compounds of lower and higher oxidation states simultaneously.

(b) Derive the half-equation for the reduction of manganate(VII) ions in acidic solution to MnO_2. Given that E^θ for this reduction is +1.67 V, show that manganese(IV) oxide will not disproportionate to MnO_4^- and Mn^{2+} in acidic solution.

$$MnO_2(s) + 4H^+(aq) + 2e^- \rightleftharpoons Mn^{2+}(aq) + 2H_2O(l), \qquad E^\theta = +1.23\ V.$$

Explanation:

$$MnO_4^-(aq) + 4H^+(aq) + 3e^- \rightleftharpoons MnO_2(s) + 2H_2O(l), \qquad E^\theta = +1.67\ V.$$

If MnO_2 disproportionates to MnO_4^- and Mn^{2+} in acidic solution, then:

$$5MnO_2(s) + 4H^+(aq) \rightarrow 3Mn^{2+}(aq) + 2H_2O(l) + 2MnO_4^-(aq).$$

The $E^\theta_{cell} = E^\theta_{Red} - E^\theta_{Ox} = 1.23 - (+1.67) = -0.44$ V.

Since $E^\theta_{cell} < 0$, the reaction is thermodynamically non-spontaneous under standard conditions; MnO_2 will not disproportionate to MnO_4^- and Mn^{2+} in acidic solution.

5. (a) Give the electronic configuration of a V^{3+} ion.

Explanation:

Electronic configuration of V atom: $1s^2 2s^2 2p^6 3s^2 3p^6 3d^3 4s^2$.
Therefore, the electronic configuration of V^{3+} is $1s^2 2s^2 2p^6 3s^2 3p^6 3d^2$.

Do you know?

— When you fill the electronic configuration till the point of atomic number 19, the energy level of the 4*s* subshell is lower than the 3*d* subshell. This is because as the atomic number increases, the energy levels of each of the subshells decrease non-proportionately as shown, due to the different penetrating powers of the different types of orbitals, which basically refers to the distance of the subshell from the nucleus:

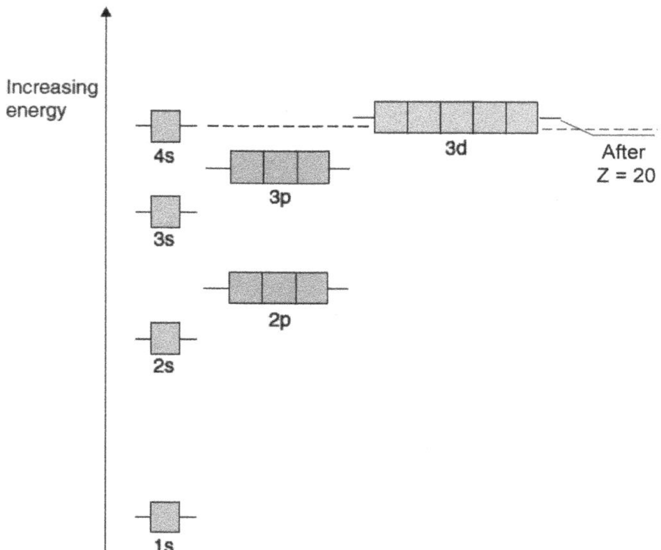

Thus, according to Aufbau's Principle, the 4*s* subshell should be filled first before the filling of the 3*d* subshell. But once we start to fill the 3*d* subshell, because the 3*d* subshell is nearer to the nucleus than the 4*s* subshell, the energy level of the 4*s* electrons is elevated. Hence, when we write down the electronic configuration of the *d* block element, it is always written with the 3*d* subshell before the 4*s* subshell.

— When we remove electrons from the atom of a transition element, we always remove the 4*s* electrons first before removing the 3*d* electrons.

Q Why are the penetrating powers of the subshells different?

A: The penetrating powers (distance from the nucleus) of the subshells in the same principal quantum shell decreases in the order: $s > p > d > f$. The corresponding energy of the subshells increases in the order: $s<p<d<f$. Across different principal quantum numbers, the relative penetrating power for the same type of subshell decreases in the order: $1s > 2s > 3s > 4s > 5s$, etc. In addition, the number of protons causes each particular subshell of a particular principal quantum number to change to a different extent. All these complex factors combine in a non-linear relationship, which coincidentally results in the $3d$ subshell's energy being higher than that of the $4s$ subshell at atomic number 19.

(b) Vanadium is a transition element. Give two characteristics of such a transition element other than the ability to form colored ions.

Explanation:

Vanadium exhibits variable oxidation states, for example: V^{2+}, V^{3+}, and VO^{2+}.

Vanadium(V) oxide, V_2O_5, is used as a catalyst in the contact process for the production of sulfuric(VI) acid from sulfur dioxide.

(c) Ammonium vanadate(V) dissolves in sulfuric acid to give a yellow solution, with the color due to the VO_2^+ ion.

(i) What is the oxidation number of vanadium in the VO_2^+ ion?

Explanation:

The oxidation number of vanadium in the VO_2^+ ion is +5.

Do you know?

— The maximum oxidation state of vanadium is +5.
— When ammonium vanadate, $(NH_4)_3VO_4$, dissolves in sulfuric (VI) acid, the following reaction takes place:

$$VO_4^{3-}(aq) + 4H^+(aq) \rightleftharpoons VO_2^+(aq) + 2H_2O(l).$$

This is an acid–base reaction, NOT a redox reaction!
— Similarly, when acid is added to yellow CrO_4^{2-}, the solution turns orange due to the formation of $Cr_2O_7^{2-}$:

$$CrO_4^{2-}(aq) + 2H^+(aq) \rightleftharpoons Cr_2O_7^{2-}(aq) + H_2O(l).$$

The above reaction is a condensation reaction or an acid-base reaction. When a base is added to the orange $Cr_2O_7^{2-}$, the position of the equilibrium shifts to the left to give back the yellow CrO_4^{2-}.

(ii) Give the systematic name of the VO_2^+ ion.

Explanation:

The systematic name of the VO_2^+ ion is vanadate(V) ion.

Do you know?

— Do not mix up VO_2^+ ion, vanadate(V) ion, with VO^{2+} which is vanadate(IV) ion.

(d) Treatment of the yellow solution from *(c)* with zinc causes the color to change to green, then to blue, followed by green again, and finally violet. Give the formulae of the ions responsible for each of these colors.

Explanation:

Yellow	Green	Blue	Green	Violet
VO_2^+	$VO_2^+ + VO^{2+}$	VO^{2+}	V^{3+}	V^{2+}

Do you know?

— Since VO_2^+ is yellow while VO^{2+} is blue, a mixture of VO_2^+ and VO^{2+} would give a green solution.

(e) In the sequence of changes in *(d)*, zinc acts as a reducing agent.

(i) State the meaning of the term *reducing agent*.

Explanation:

A reducing agent decreases the oxidation state of another species while itself is being oxidized through losing electrons. Hence, the oxidation number of the reducing agent increases.

(ii) Write a half-equation showing how zinc acts as a reducing agent.

Explanation:

Oxidation half-equation: $Zn \rightarrow Zn^{2+} + 2e^-$.

(iii) Write the half-equation for the conversion of VO_2^+ to VO^{2+} in acidic solution.

Explanation:

Reduction half-equation: $VO_2^+ + 2H^+ + e^- \rightarrow VO^{2+} + H_2O$.

(iv) Hence, write the equation for the reduction of VO_2^+ to VO^{2+} by zinc.

Explanation:

Overall redox reaction: $2VO_2^+ + 4H^+ + Zn \rightarrow 2VO^{2+} + 2H_2O + Zn^{2+}$.

Do you know?

— For the reduction of VO_2^+ by Zn:

$$VO_2^+ + 2H^+ + e^- \rightleftharpoons VO^{2+} + H_2O, \qquad E^\theta = +1.00\text{ V, and}$$
$$Zn^{2+} + 2e^- \rightleftharpoons Zn, \qquad E^\theta = -0.76\text{ V.}$$
$$2VO_2^+ + 4H^+ + Zn \rightarrow 2VO^{2+} + 2H_2O + Zn^{2+}.$$

The $E^\theta_{cell} = E^\theta_{Red} - E^\theta_{Ox} = 1.00 - (-0.76) = +1.76$ V.

Since $E^\theta_{cell} > 0$, the reaction is thermodynamically spontaneous under standard conditions.

— For the reduction of VO^{2+} by Zn:

$$VO^{2+} + 2H^+ + e^- \rightleftharpoons V^{3+} + H_2O, \qquad E^\theta = +0.34\text{ V, and}$$
$$Zn^{2+} + 2e^- \rightleftharpoons Zn, \qquad E^\theta = -0.76\text{ V.}$$
$$2VO^{2+} + 4H^+ + Zn \rightarrow 2V^{3+} + 2H_2O + Zn^{2+}.$$

The $E^\theta_{cell} = E^\theta_{Red} - E^\theta_{Ox} = 0.34 - (-0.76) = +1.10$ V.

Since $E^\theta_{cell} > 0$, the reaction is thermodynamically spontaneous under standard conditions.

— For the reduction of V^{3+} by Zn:

$$V^{3+} + e^- \rightleftharpoons V^{2+}, \qquad E^\theta = -0.26\text{ V, and}$$
$$Zn^{2+} + 2e^- \rightleftharpoons Zn, \qquad E^\theta = -0.76\text{ V.}$$
$$2V^{3+} + Zn \rightarrow 2V^{2+} + Zn^{2+}.$$

The $E^\theta_{cell} = E^\theta_{Red} - E^\theta_{Ox} = -0.26 - (-0.76) = +0.50$ V.

Since $E^\theta_{cell} > 0$, the reaction is thermodynamically spontaneous under standard conditions.

(Continued)

(*Continued*)

— For the reduction of V^{2+} by Zn:

$V^{2+} + 2e^- \rightleftharpoons V$, $E^\theta = -1.20$ V, and
$Zn^{2+} + 2e^- \rightleftharpoons Zn$, $E^\theta = -0.76$ V.
$V^{2+} + Zn \rightarrow 2V + Zn^{2+}$.

The $E^\theta_{cell} = E^\theta_{Red} - E^\theta_{Ox} = -1.20 - (-0.76) = -0.44$ V.

Since $E^\theta_{cell} < 0$, the reaction is thermodynamically non-spontaneous under standard conditions. Hence, reduction by Zn will stop at the V^{2+} state!

Q What happen when the $V^{2+}(aq)$ is left exposed to air?

A: The V^{2+} will be slowly oxidized by O_2 back to V^{3+}, then to VO^{2+}, and then to VO_2^+.

6. Coins are made from an alloy, nickel–brass, which consists essentially of the metals copper, nickel, and zinc. A one-pound coin weighing 9.50 g is completely dissolved in concentrated nitric acid, in which all three metals dissolve, to give solution **A**.

   Dilute sodium hydroxide solution was then added carefully with stirring, until present in excess. Zinc hydroxide is amphoteric. The precipitate formed, **B**, is filtered off from the supernatant liquid, **C**. The precipitate, **B**, is quantitatively transferred to a graduated flask of 500 cm^3 capacity. Dilute sulfuric acid is then added dropwise to dissolve the whole of precipitate **B** and the solution was made up to 500 cm^3 with distilled water.

   25.0 cm^3 of this solution were pipetted into a conical flask and an excess of potassium iodide solution was added. The liberated iodine then titrated against a sodium thiosulfate solution of concentration 0.100 mol dm^{-3}. 18.7 cm^3 of the sodium thiosulfate solution was required for a complete reaction.

   (a) (i) With reference to the Data Booklet, using the appropriate half-equations, write an equation for the reaction of any one of the metals in nickel–brass with concentrated nitric acid.

Explanation:

$$Zn^{2+} + 2e^- \rightleftharpoons Zn \qquad E^\theta = -0.76 \text{ V, and}$$
$$2H^+ + 2e^- \rightleftharpoons H_2, \qquad E^\theta = 0.00 \text{ V.}$$

The $E^\theta_{cell} = E^\theta_{Red} - E^\theta_{Ox} = 0.00- (-0.76) = +0.76$ V.

Since $E^\theta_{cell} > 0$, the reaction is thermodynamically spontaneous under standard conditions. Therefore, the overall redox reaction is:

$$Zn + 2H^+ \rightarrow Zn^{2+} + H_2.$$

Do you know?

— For nickel:

$$Ni^{2+} + 2e^- \rightleftharpoons Ni, \qquad E^\theta = -0.25 \text{ V, and}$$
$$2H^+ + 2e^- \rightleftharpoons H_2, \qquad E^\theta = 0.00 \text{ V.}$$

The $E^\theta_{cell} = E^\theta_{Red} - E^\theta_{Ox} = 0.00 - (-0.25) = +0.25$ V.

Since $E^\theta_{cell} > 0$, the reaction is thermodynamically spontaneous under standard conditions. Therefore, the overall redox reaction is:

$$Ni + 2H^+ \rightarrow Ni^{2+} + H_2.$$

— For copper:

$$Cu^{2+} + 2e^- \rightleftharpoons Cu, \qquad E^\theta = +0.34 \text{ V, and}$$
$$2H^+ + 2e^- \rightleftharpoons H_2, \qquad E^\theta = 0.00 \text{ V.}$$

The $E^\theta_{cell} = E^\theta_{Red} - E^\theta_{Ox} = 0.00- (+0.34) = -0.34$ V.

Since $E^\theta_{cell} < 0$, the reaction is thermodynamically non-spontaneous under standard conditions. Therefore, the Cu metal cannot react with H^+. But then why did the Cu metal dissolve in concentrated HNO_3?

$$Cu^{2+} + 2e^- \rightleftharpoons Cu, \qquad E^\theta = +0.34 \text{ V, and}$$
$$NO_3^- + 2H^+ + e^- \rightleftharpoons NO_2 + H_2O, \qquad E^\theta = +0.81 \text{ V.}$$

The $E^\theta_{cell} = E^\theta_{Red} - E^\theta_{Ox} = 0.81- (+0.34) = +0.47$ V.

Since $E^\theta_{cell} > 0$, the reaction is thermodynamically spontaneous under standard conditions. Therefore, the Cu metal reacts with the NO_3^- instead.

Q So, copper metal would not react with HCl(aq) or H_2SO_4(aq)?

A: Yes, there would not be any reaction between Cu(s) and the components in HCl(aq) or H_2SO_4(aq).

Q Can Cu(s) reduce NO_3^- to HNO_2 or NH_4^+?

A: We can calculate the E^θ_{cell} values for the following reactions:

$$Cu^{2+} + 2e^- \rightleftharpoons Cu, \qquad E^\theta = +0.34 \text{ V, and}$$
$$NO_3^- + 3H^+ + 2e^- \rightleftharpoons HNO_2 + H_2O, \qquad E^\theta = +0.94 \text{ V.}$$

The $E^\theta_{cell} = E^\theta_{Red} - E^\theta_{Ox} = 0.94 - (+0.34) = +0.60$ V.

Since $E^\theta_{cell} > 0$, the reaction is thermodynamically spontaneous under standard conditions. Cu metal can reduce NO_3^- to HNO_2.

$$Cu^{2+} + 2e^- \rightleftharpoons Cu, \qquad E^\theta = +0.34 \text{ V, and}$$
$$NO_3^- + 10H^+ + 8e^- \rightleftharpoons NH_4^+ + 3H_2O, \qquad E^\theta = +0.87 \text{ V.}$$

The $E^\theta_{cell} = E^\theta_{Red} - E^\theta_{Ox} = 0.87 - (+0.34) = +0.53$ V.

Since $E^\theta_{cell} > 0$, the reaction is thermodynamically spontaneous under standard conditions. Therefore, the Cu metal can reduce NO_3^- to NH_4^+.

Q To which product would the reduction of NO_3^- by Cu metal be more probable, NO_2, HNO_2, or NH_4^+?

A: We need more H^+ ions for the reduction of NO_3^- to HNO_2 or NH_4^+, hence reducing NO_3^- to NO_2 is more probable. This fits with our observation in that when we add a piece of copper metal to HNO_3, the brown gas of NO_2 is seen.

(ii) What type of reaction is taking place?

Explanation:

A redox reaction is taking place.

(b) Identify by giving full formulae:

(i) the complex cations present in A;

Explanation:

The complex cations present in **A** are $[Cu(H_2O)_6]^{2+}$, $[Ni(H_2O)_6]^{2+}$, and $[Zn(H_2O)_6]^{2+}$.

Q Why does $Zn(OH)_2$ dissolve in excess NaOH(aq)?

A: When NaOH(aq) is added to $[Zn(H_2O)_6]^{2+}$, the ionic product for the formation of $Zn(OH)_2(s)$ exceeds its K_{sp} value, hence the white ppt of $Zn(OH)_2$ is formed:

$$[Zn(H_2O)_6]^{2+}(aq) + 2OH^-(aq) \rightleftharpoons [Zn(OH)_2(H_2O)_4](s) + 2H_2O(l).$$

When excess NaOH(aq) is added, $[Zn(OH)_4]^{2-}$ is formed due to ligand exchange:

$$[Zn(H_2O)_6]^{2+}(aq) + 4OH^-(aq) \rightleftharpoons [Zn(OH)_4]^{2-}(aq) + 6H_2O(l).$$

According to Le Chatelier's Principle, $[Zn^{2+}(aq)]$ decreases, position of equilibrium shift left, hence more $Zn(OH)_2$ dissolved.

(ii) the precipitates in **B**; and

Explanation:

The precipitates in **B** are $Cu(OH)_2$ and $Ni(OH)_2$.

(iii) any metal-containing anion in **C**.

Explanation:

The metal-containing anion in **C** is $[Zn(OH)_4]^{2-}$.

(c) (i) Write an equation for the precipitate of any one of the metal ions in **A** with sodium hydroxide.

Explanation:

$$[Zn(H_2O)_6]^{2+}(aq) + 2OH^-(aq) \rightarrow [Zn(OH)_2(H_2O)_4](s) + 2H_2O(l).$$

Q Why did you use a single arrow '→' instead of a double arrow?

A: We are trying to show that a reaction has occurred, although in reality, there is an equilibrium being established. A double-arrow equation is used if we want to make use of that equation to explain observations based on equilibrium shift.

(ii) What type of reaction is occurring in *c(i)*?

Explanation:

The reaction that occurred in *c(i)* is a ligand exchange reaction.

Q Can we use the term 'precipitation' here?

A: Try not to as in transition metal chemistry, it is normally perceived as 'ligand exchange.' Under the K_{sp} section, we view it as precipitation reaction. Actually, it all depends on who the examiner is and what he/she wants.

(d) Suggest an explanation on why it is necessary to add sodium hydroxide, followed by dilute sulfuric acid, before performing the titration.

Explanation:

The sodium hydroxide is used to isolate the Cu^{2+} from its other impurities as insoluble solid, $Cu(OH)_2$. The sulfuric(VI) acid is used to dissolve $Cu(OH)_2$ into the solution form, so that we can carry out our titration to determine the amount of Cu^{2+}.

(e) On addition of the potassium iodide solution, the only reaction which occurs is:

$$2Cu^{2+}(aq) + 4I^{-}(aq) \rightarrow 2CuI(s) + I_2(aq).$$

(i) Write an equation for the reaction between sodium thiosulfate and the liberated iodine. What indicator would you use in this titration? At what stage would you add it? Give a reason for your answer.

Explanation:

The reaction is: $I_2(aq) + 2S_2O_3^{2-}(aq) \rightarrow 2I^{-}(aq) + S_4O_6^{2-}(aq)$.

Starch solution as an indicator would be added toward the end of titration when the solution is pale yellow in color. This is because when starch solution added, a blue-black starch–iodine complex would form. When more sodium thiosulfate is added, the color transition from blue-black to colorless is more prominent than from pale yellow to colorless.

(ii) Calculate the percentage of copper in the alloy.

Explanation:

Amount of $S_2O_3^{2-}(aq)$ used = $0.0187 \times 0.100 = 1.87 \times 10^{-3}$mol.

$$I_2(aq) + 2S_2O_3^{2-}(aq) \rightarrow 2I^{-}(aq) + S_4O_6^{2-}(aq).$$

Amount of I_2 = ½ × Amount of $S_2O_3^{2-}$(aq) used = 9.35×10^{-4} mol.

$$2Cu^{2+}(aq) + 4I^-(aq) \rightarrow 2CuI(s) + I_2(aq).$$

Amount of Cu^{2+}(aq) in 25 cm^3 = 2 × Amount of I_2 = 1.87×10^{-3}mol.
Amount of Cu^{2+}(aq) in 500 cm^3 =20 × 1.87×10^{-3} = 0.0374 mol.
Mass of Cu in 9.50 g of alloy = 0.0374 × 63.5 = 2.37 g.
Percentage of copper in the alloy $= \frac{2.37}{9.50} \times 100 = 25.0\%$.

(iii) Suggest why this reaction occurs in the light of the E^θ values from the Data Booklet.

Explanation:

$$Cu^{2+} + e^- \rightleftharpoons Cu^+, \quad E^\theta = +0.15 \text{ V, and}$$
$$I_2 + 2e^- \rightleftharpoons 2I^-, \quad E^\theta = +0.54 \text{ V.}$$

The E^θ_{cell} calculation for $2Cu^{2+}(aq) + 4I^-(aq) \rightarrow 2CuI(s) + I_2(aq)$ is:
$E^\theta_{cell} = E^\theta_{Red} - E^\theta_{Ox} = 0.15- (+0.54) = -0.39$ V.

Since $E^\theta_{cell} < 0$, the reaction is thermodynamically non-spontaneous under standard conditions. But the reaction proceeds because the CuI formed is insoluble in water, hence the removal of the Cu^+ from the reaction drives the reaction forward.

Do you Know

— Cu^+ is highly unstable in water and undergoes disproportionation readily to give a blue solution of Cu^{2+} and the pink Cu metal:

$$Cu^{2+} + e^- \rightleftharpoons Cu^+, \quad E^\theta = +0.15 \text{ V, and}$$
$$Cu^+ + e^- \rightleftharpoons Cu, \quad E^\theta = +0.52 \text{ V.}$$

The E^θ_{cell} calculation for $2Cu^+(aq) \rightarrow Cu(s) + Cu^{2+}(aq)$ is:

$$E^\theta_{cell} = E^\theta_{Red} - E^\theta_{Ox} = 0.52 - (+0.15) = +0.37 \text{ V.}$$

Since $E^\theta_{cell} > 0$, the reaction is thermodynamically spontaneous under standard conditions.

INDEX

www.ingramcontent.com/pod-product-compliance
Lightning Source LLC
Chambersburg PA
CBHW050539190326
41458CB00007B/1838